Public Libraries, Institutions and Societies

Public Libraries, Institutions and Societies

Editor

Ralph Wagner

scitus academics

Public Libraries, Institutions and Societies
Edited by **Ralph Wagner**

ISBN: 978-1-68117-634-5
Library of Congress Control Number: 2016946494

SCITUS Academics LLC,
616, Corporate Way, Suite 2, 4766,
Valley Cottage, NY 10989

www.scitusacademics.com

Notice

Printed in United States of America on Acid Free Paper

Preface

The Public Library is regarded in the people's institution. It is local gateway to knowledge that provides basic condition for life learning and facilitates cultural development of the individual and social groups. The public library provides services to all irrespective of age, sex, cast, religion, education, and social studies, generally all users get the material relevant to their needs and requirements in the public library. The advent of information technology from the middle of 20th century has gradually changed the traditional process of information acquisition, communication and storage. The society is transformed with new a paradigm which is regarded as information society or knowledge society. Along with the traditional form of reading material such as books, manuscripts the new forms like microfiche, CDs, e-books, etc. In a society of lifelong learning - whether of a formal or informal nature- public libraries will be nodes connecting the local learning setting with the global resources of information and knowledge. Public libraries must therefore be allowed to play a role of fundamental importance in the development of future systems of lifelong learning. The development of the information and communication technology has already laid the basis for the creation of information networks, giving users even of small local public libraries access to the world wide sources of information. The public libraries also offer guidance and training in information search and quality rating of information sources. Thus, public libraries are necessary if not sufficient prerequisites for a democratic knowledge society. The importance of the idea of the library organization as a social institution became so powerful that it also became a basic tenet of the library field, the basis for what since that time has become one of the field's common assumptions, that the modern library in and of itself, the library as a social organization, is the field's most significant cultural legacy to society.

Libraries and other similar type of institutions collect, process, organize and disseminate information and knowledge recorded in document. Since knowledge and information are vital for all round development of human beings, libraries and other institutions that handle and manage information and knowledge are indeed invaluable for the welfare of the society.

Public Libraries, Institutions and Societies is an immense guide for librarians, researchers and practitioners.

TABLE OF CONTENTS

1

CHAPTER

The Impacts of Economic Development on Public Libraries: A Study Case of Bibliotheca Alexandrina Library

Samaa Soliman, Pan Wei

China Agricultural University Library, Beijing, China

1. INTRODUCTION

The Public Library is the local center of information, making all kinds of knowledge and information readily available to its users. Besides, it considered the local gateway to knowledge, and provided a basic condition for lifelong learning, independent decision-making and cultural development of the individual and social groups. As the Public Library shall in principle be free of charge. The public library is the responsibility of local and national authorities. It must be supported by specific legislation and financed by national and local governments [1] . Therefore, public libraries in most cases are affected directly by any decisions made by local and national authorities concerning its budget. And these decisions obviously related to both local and nation-wide economy development.

This research is focused on empirical evidence to clarify the impact of economic development on the public libraries, where it is trying to measure these impacts using quantitative analysis based on historical data from Bibliotheca Alexandrina Library.

Researchers have long been concerned with the relation between economic development and public libraries, while a lot of studies were conducted to examine the role of public libraries in economic development; only a few studies investigated the impacts of economic development on public libraries.

As an example for the first group, a research has been conducted on Northport-East Northport Public Library, which estimates the economic impact of the operating and capital expenditures made by the library. And one of the key findings was that the library's operating and capital expenditures resulted in increased output, earnings and employment throughout Long Island but particularly in its local community [2] . Another study was conducted to measure the economic benefits of Indiana's public libraries, which found that Libraries are a good value. The direct economic benefits that communities receive from libraries are significantly greater than the cost to operate the libraries [3] . The Institute for Public Policy and Business Research undertook a study to determine how Kansas public libraries could play a more active role in supporting business and economic development in their communities. The major findings show that those libraries play an important role in providing materials and services to the business community [4] . Another study aimed to measure the value of Norwegian public library benefits to the population in monetary terms, which conclude that the Norwegian public libraries are worth four times more than they cost [5] . A study discussed the public library's role in society as a focal point for local economic development, which promotes the idea of setting up an information service in public libraries to provide added value information; which will be of great importance in justifying the existence of the public library within its surroundings [6] .

Similar results showed that Toronto Public Library delivers a strong return on investment, through the delivery of library services, as it showed that for every dollar invested in Toronto Public Library (TPL), Torontonians receive $5.63 [7] . Trying to determine economic benefit of public libraries for South Carolinians and to what extent they feel the public library contributes to their overall economic wellbeing. Results showed that the total direct economic impact of SC public libraries is estimated at $222 million, while the actual cost of these services to the state and local governments is only $77.5 million. This means that for every $1 spent by state and local governments on SC public libraries the return on investment is $2.86 [8] .

From even boarder prospective, another study found that the return on investment in public libraries not only benefits individuals, but also strengthens community capacity to address urgent issues related to economic development. Public libraries are increasingly finding their "fit" in the formal and informal network of agencies, corporations, nonprofits, and community organizations working together to elevate levels of education and economic potential, making cities stronger [9] . Lastly, a study was conducted on selected Colorado Public Libraries in May 2006. Using a multiple case study approach, it is found that for most of the libraries participating in the study, the Return on Investment (ROI) was approximately five to one—that is, for every $1.00 spent on public libraries, $5.00 of value was realized by taxpayers [10] .

On the other hand, as an example for the second group, other studies and reports investigated the impact of economic development on public libraries. The virtual entirety of public libraries in the US and UK were then, and still being, dependent on local or central public finance. These major sources of revenue were constrained as a result of these political and economic changes because the reduction of public spending was one of the major goals advocated [11] . The public library has no major source of income other than local tax revenue. Income from gifts and endowments was only 4.3 percent of the total in 1956 [12] . Other research tried to point out the importance of learning from the past as Libraries should take action and be proactive in the response to the challenges, that lay before them, they should publicize the importance of the work libraries do and the services they provide, and finally libraries should cooperate with other agencies who are providing other social and government services [13] . Another study emphasized the role of public libraries and how this role has significantly increased and libraries have greatly contributed to the improvement of information society, giving that there is a strong statistically significant linear correlation between the percentage of Internet users and the natural logarithm of GDP, which confirms, that the usage of ICT is one of the main drivers of change within society and economic development [14] . A research focusing primarily on acquisitions claimed that the libraries' main funding source is the national government; thus, their fortunes are tied to the broader economy. Libraries are especially concerned about paying for electronic resources, given their importance and high cost [15] . Another study suggests that in response to recent public library funding deficiencies, many people both inside and outside the field of librarianship have suggested that public libraries need to rely less on traditional government funding and more on alternative funding sources [16] .

2. DATA SOURCES & METHODOLOGY

The data has been used in this research is based on a secondary data collected from various BA annual reports, mainly from 2003 until 2012, as for the data concerning the Egyptian economy, it has been extracted mainly from world bank data. Then, a data descriptive analysis is implemented to summarize the data.

To answer the research question about the impact of economic development on the BA library, a Pearson product-moment correlation coefficient is evaluated on two steps; in the first step, the relationship between the GDP (PPP)—as one important economic development indicator, and BA allocated budget is investigated.

$$r = \frac{n\left(\sum \text{GDP} * B\right) - \left(\sum \text{GDP}\right)\left(\sum B\right)}{\sqrt{\left[n\sum \text{GDP}^2 - \left(\sum \text{GDP}\right)^2\right] + \left[n\sum B^2 - \left(\sum B\right)^2\right]}} \cdots \text{where } B \text{ is the Allocated Budget.}$$

Then in the second step, various library indicators are tested against the allocated budget such as the different library collection categories, membership, visitors, salaries & wages, and the general expenses as Figure 1 illustrates.

3. GDP GROWTH AND THE LIBRARY INVESTMENT INCREASE

The Egyptian economy is one of the Middle East's most versatile economies, which the sectors of agriculture, industry, tourism and services engaged in comparable proportions in its basic configuration. The average number of the work force in Egypt is about 26 million, according to 2010 estimates. They are distributed on the service sector (51%), agriculture sector (32%) and the industrial sector (17%). The Egyptian economy depends basically on agriculture, Suez Canal revenues, tourism, taxation, cultural and media production, petroleum exports and remittances of more than three million Egyptians abroad, mostly in the Gulf States, the United States, Europe and Australia [17] .

Egypt has one of the most developed and diversified economies in the Middle East. Until 2010, Egyptian economy was growing an average 5 percent a quarter as a result of several economic reforms attracting foreign investments. During that time, the economy and the living standards for majority of population improved. Yet, living conditions for the average Egyptian still remained poor and large income disparities continued to grow, leading to the public discontent. In the wake of the January 25th Revolution, the country exerted strenuous efforts to restore the economy on right track and overcome the obstacles through rallying the efforts internally and openness to the world to act dynamically in the world's economy and finance. And also as it brought down President Hosni Mubarak regime, have caused economic slowdown as political and institutional uncertainty and rising insecurity continue to hurt tourism, manufacturing, and construction [18] .

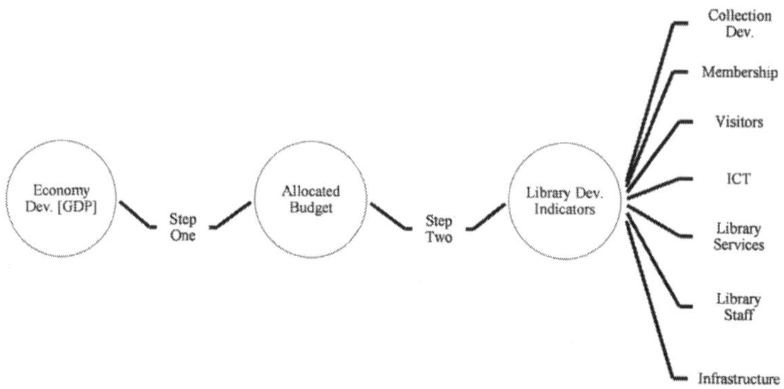

Figure 1. Analysis methodology.

One important indicator which is the GDP PPP, which shows in general increase in the country GDP from $565 billion in 2003 to reach about $862 Billion in 2012 as Figure 2 shows [19] .

On the other hand, the Egyptian public library movement, which had already acquired momentum before 1952, benefited from Nasser's support for mass education during the first decade of the Egyptian revolution. The perception that education fosters national development induced the new regime to take deliberate actions in support of the library movement in Egypt, in the early stages of the revolution. School libraries were given special attention as part of the government's overall interest in mass education, which the intention of making school libraries "an integral part of school education". Government and special libraries were also among the Nasser regime's highest priorities. Convinced of the need for scientific information in research, teaching and industry, the new regime established new science and research libraries. The Egyptian public library movement, especially the Egyptian National Library "Dar al-Kutub" (House of Books), benefited directly from Nasser's early affirmation of education as a birthright of every Egyptian. Annual book acquisitions almost tripled between 1952/53 and 1961/62. The spread of education and culture was also associated with a significant increase in Egyptian book production, and the Copyright-Legal Deposit Law enacted by Nasser's regime in 1954 guaranteed Dar al-Kutub and its branch public libraries a constant acquisition of national publications. Moreover, a total of sixty popular culture libraries were established in rural Egypt between 1958 and 1970. Then, the number of public libraries, library users, book acquisitions, and national library budgets continued to expand—with not the same speed as before -during the Sadat era (1970-1981) [20] and Mubarak era as well (1982-2011).

Libraries are located throughout Egypt, including remote regions in Upper Egypt and Sinai. As expected, most of the libraries are located in greater Cairo and Alexandria. Nasser's dream of a library in every village, regardless of size, has not been achieved.

It can be seen that the per capita number of libraries in Egypt is uneven with respect to county or governorate. The governorate with the lowest number of people/library is Suez (3,440 people per library) and the lowest village is Sohag (118,149 people per library). A number of governorates—all in Upper Egypt—stand out with the fewest libraries per capita. They are Beni Sweif, Qena, Behaira and Sohag as Table 1 shows.

The average per capita library figure in the United States is approximately 17,000. This means that there are about twice as many libraries per capita in the US compared to Egypt.

The rate of new library construction lags population growth in Egypt. The projected population growth in Egypt in 1.61% [19] compared with new construction for libraries at less than 0.90%. The rate of new library construction should be increased, especially in places such as the Fayoum that have not had a new library in built in nearly 10 years and have a rapidly growing population as Figure 3 shows.

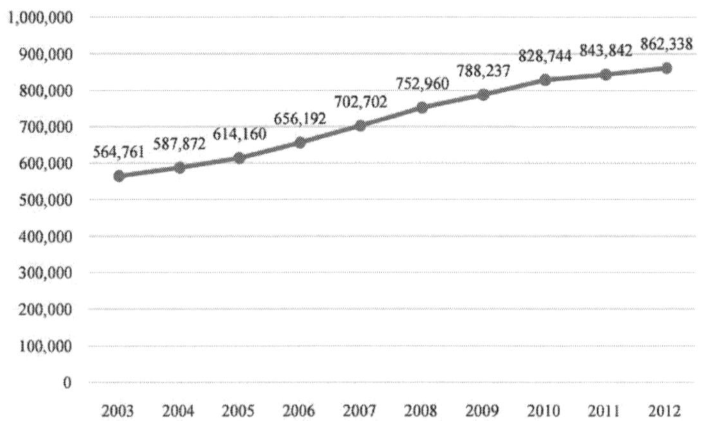

Figure 2. Egypt GDP PPP (constant 2011 international $), billion. Source: [19].

Table 1. Libraries per capita in Egypt.

Governorates	Total population	Poverty rate	No of libraries	Persons/library
Al Wadi Al Gaded	169,896	6.24	42	4045
Alexandria	4,087,682	6.41	73	55,996
Assuit	2,953,561	60.97	124	23,819
Aswan	1,187,943	40.92	33	35,998
Beni Sweif	2,253,076	41.45	26	86,657
Behaira	4,676,382	23.51	78	59,954
Cairo	7,774,143	7.62	308	25,241
Dakahleya	4,830,513	9.30	256	18,869
Domiat	1,074,948	1.11	64	16,796
Fayoum	2,441,333	28.71	44	55,485
Gharbya	3,940,061	7.64	59	66,781
Giza	6,562,923	22.97	176	37,289
Ismailia	832,315	18.84	41	20,300
Kafr Al Shikh	2,576,792	11.20	43	59,925
Ismailia	832,315	18.84	41	20,300
Kafr Al Shikh	2,576,792	11.20	43	59,925
Luxor	440,110	18.44	10	44,011
Marsa Matrouh	212,219	4.32	8	26,527
Menia	2,895,597	30.93	36	80,433
Menofia	3,941,818	17.93	56	70,390
North Sinai	300,502	27.87	22	13,659
Port Said	384,571	4.43	40	9614
Qalyoubya	4,175,659	11.33	111	37,619
Qena	2,938,079	39.02	32	91,815
Red Sea	225,937	4.00	10	22,594
Sharkeya	5,307,846	19.15	71	74,758
Sohag	3,662,619	47.54	31	118,149
South Sinai	116,379	0.00	10	11,638
Suez	103,478	1.94	30	3449
Total	70,066,382	21.56	1834	38,204

A large number of private or semi-private libraries exist at universities, schools, social, sport and military clubs and in mobile libraries. These collections should be considered when thinking of the overall access to libraries in Egypt [21] .

However, in general the number of public libraries in Egypt fluctuated over time, as it reached its minimum value in 2003 with only 112 thousand visitors, and reached its maximum value in 2013 with 602 thousand visitors [22] as Figure 4 illustrates.

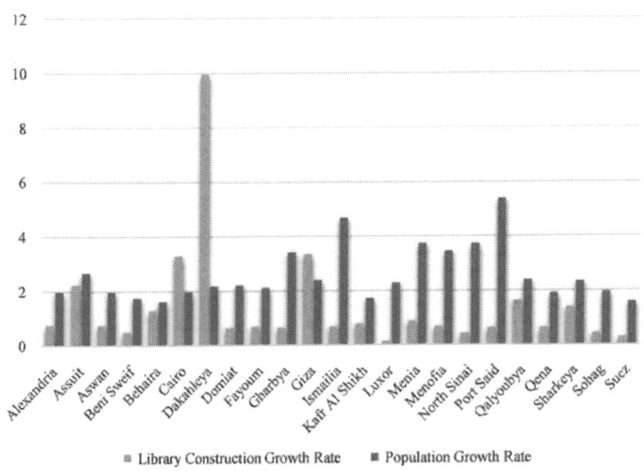

Figure 3. Population growth rate and library construction growth rate in different Egyptian governorates.

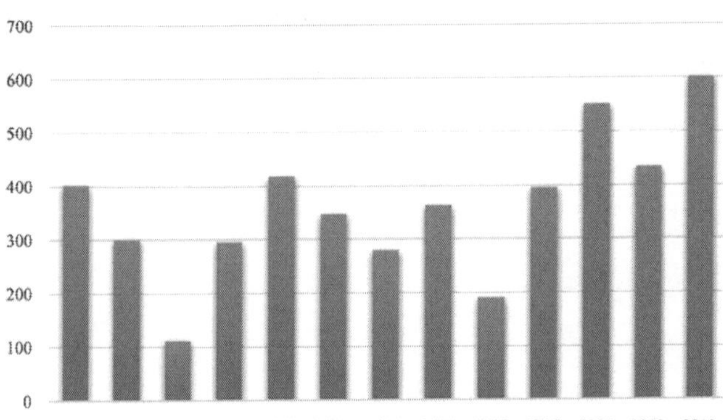

Figure 4. Number of libraries visitors in thousands. Source: [22] .

4. THE CASE STUDY OF BIBLIOTHECA ALEXANDRINA

The idea of reviving the ancient library of Alexandria had been floating around since 1974, when a committee set up by the Alexandria University selected a plot of land for its new library, between the campus and the seafront, close to where the ancient library once stood. The notion of recreating the ancient library was soon enthusiastically adopted by other individuals and agencies throughout Egypt. The United Nations Educational, Scientific, and Cultural Organization (UNESCO) was quick to embrace the idea of endowing the Mediterranean region with a center of cultural and scientific excellence, and the project was well underway by the late 1980s. In 1990 a conference was held in Aswan to help the government figure out the logistics of funding such as costly project. 65 million dollars was pledged b various agencies, mainly from the gulf Arab states. Construction on the library's complex began in 1995 by Snøhetta, a Norwegian architectural firm. After two-hundred and twenty million dollars in total was spent, the Bibliotheca Alexandria officially opened on 16th, 2002 [23] .

The library has shelf space for eight million books, with the main reading room covering 20,000 square meters on eleven cascading levels as displayed in Figure 5. The complex also houses a conference center; specialized libraries for maps, multimedia, the blind and visually impaired, young people, and for children; four museums; four art galleries for temporary exhibitions; 15 permanent exhibitions; a planetarium; and a manuscript restoration laboratory. The library's architecture is equally striking. The main reading room stands beneath a 32-meter-high glass-paneled roof, tilted out toward the sea like a sundial, and measuring some 160 m in diameter. The walls are of gray Aswan granite, carved with characters from 120 different human scripts [24] .

4.1. Allocation of Budget and Donations

Bibliotheca Alexandrina have its own independent budget, and its financial year starts and end with the financial year of the government's budget. And the Library have a separate account in the Central Bank of Egypt as well. In general, the library has the following resources of funding:

- The allocations provided for it by the state.
- The support, gifts, donations, bequests, and financial contributions from internal or external sources.
- The income that it receives for its services and the returns on the invested funds in its projects.
- Any other sources that may be legally allocated to the Library [25] .

Since Aswan meeting, the library receives a lot of donations and contributions from many countries, international agencies and companies such as: United States Agency for International Development (USAID), Qatar Emirate, European Commission (EC), World Bank, UNESCO, Islamic Development Bank (IDB), and others.

Table 2 shows how the allocation of budget changes over time—which mainly represents grants from the ministry of finance against wages and salaries, and current expenses. As the government increases the budget allocated to the library from around 17 million EGP in 2003 which represents only about 31% from the total revenues, to reach about 125 million EGP in 2012, which represents around 75% of the total revenues. On the other hand, the other sources of revenues such as donations and the library own investment revenue and projects, which was represents about 69% of total revenues decreased to become 25% only of the total revenues. As an average, the government allocated budget represents around 83 million EGP of the library revenues comparing to about 40 million EGP of revenues from donations & projects as Figure 6 shows.

As for the expenses, there are two main categories of it, one is salaries & wages, which was around 20 million EGP in 2003 and reached about 108 million EGP in 2012. The other category is the general and administrative expenses, that represents all operation costs such as conference expenses, utilities, exhibitions, taxes, travel, etc. which increased from 29 million EGP in 2003 to reach about 48 million EGP in 2012. Other category of expenses represents provisions providing for employees' end-of-service benefits, contingent liabilities and allowance for bad debt.

Figure 5. Interior view of the library.

Table 2. Revenues & expenses of BA—million EGP.

Year	Revenues					Expenses						
	Gov. grants	%	Donation & projects	%	Total	Salaries & wages	%	General expenses	%	Other*	%	Total
2003	17.21	31.01	38.29	68.99	55.50	19.81	24.54	29.06	36.00	31.85	39.46	80.72
2005	59.55	70.72	24.65	29.28	84.20	34.21	42.53	45.23	56.23	1.00	1.24	80.44
2006	55.46	62.74	32.93	37.26	88.39	38.10	44.34	47.83	55.66	0.00	0.00	85.93
2007	80.64	73.37	29.26	26.63	109.90	50.99	48.08	53.06	50.03	2.00	1.89	106.06
2008	76.97	63.10	45.01	36.90	121.99	61.20	49.19	61.73	49.61	1.50	1.21	124.44
2009	95.03	67.98	44.77	32.02	139.80	74.35	54.80	58.82	43.35	2.50	1.84	135.66
2010	110.95	66.96	54.75	33.04	165.70	90.42	56.27	66.76	41.55	3.50	2.18	160.67
2011	129.18	73.24	47.21	26.76	176.39	113.09	64.73	57.95	33.17	3.66	2.09	174.71
2012	124.89	75.18	41.23	24.82	166.11	108.33	66.09	48.39	29.52	7.21	4.40	163.93

Other*: includes Provisions & Depreciation for the first year. Source: [26].

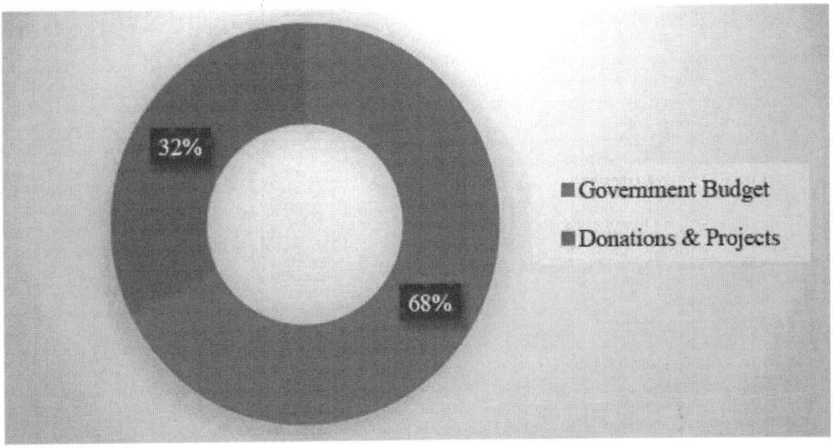

Figure 6. Source of revenues of BA. Source: [26] .

In 2012, the government underfunded the salaries budget below the level of actual expenditure of the financial year 2011. So, to confront the situation, the management tightened spending, refused to replace outgoing staff, allowing attrition to reduce the wage bill, along with a general tightening of expenditures. But the reduction in activities also meant a reduction in revenues. The budget, which was successfully balanced at the end of the year but did not allow for any salary increases to anyone, not even to cover inflation. The team of BA also arranged for a cut of 10% in the top salaries which enabled the management to give the lower salary levels 5% increases, within the same frozen, underfunded envelope [27] .

4.2. Library Components

The Library of Alexandria complex includes: The Main Library, the Taha Hussein Library for the visually impaired, the Young People Library, the Children Library, Multimedia Library, Antiquities Museum, Manuscript Museum, Science Museum,

Planetarium, International School of Information Studies (ISIS), Calligraphy and Script Institute, Conservation and Restoration Laboratory, Center Documentation and Research, Conference Center, ancillary services, multipurpose rooms and exhibition areas.

4.3. Electronic Resources

The Information and Communication Technology (ICT) infrastructure is arguably the best in Egypt: Fiber Optic backbone and Gigabit Ethernet technology throughout our buildings with 155 Mbps Internet connectivity; with over 360 public workstations and 940 for the staff, with over 60 servers. Most of the standard functions have been computerized: Library Information System, integrated digital library services, intranet, Enterprise Resource Planning and access control and ticketing systems [28] . The library is also offering a selection of databases and websites to serve all users and researchers in most areas of specialization with e-books and full text online for periodicals or other research tools. Internet Archive contributed through the generosity of Brewster Kahle, its founder and CEO, is a backup copy of material available on the Internet from 1996 to the present, much of which is no longer on the Internet. It contains over 100 terabytes of data (which is equivalent to about 10 billion pages, 2000 hours of video news films and 1000 documentary films. The material from the Internet represents snapshots at different times of the topics from over 16 million sites representing the voices of millions of authors and web initiators over the last six years.

4.4. Library Services

Reception & Visitors Services: to guide new comers and explain to visitors the historical background. It also, gives tours to explain the services and activities of the Library and other sections of the cultural complex.

Information Services: Open Access Shelves display the library's collections through a reading area, which cascades over the 7 levels of the library, and accommodates 2000 readers at the same time.

The Bibliotheca Alexandrina is thought to have the largest reading room of all the major libraries of the world. It also serves researchers with 200 study rooms spread over the 7 levels of the library. Each information desk is designed to serve the public and acquaint the user or researcher with OPAC, electronic resources, the multimedia and collections and other facilities.

General Information Services: The General Information Services desk is located at the Entrance of the Main Library. It gives general information to the public and refer users to the appropriate specialized information desk.

Orientation Services: Orientation programs are scheduled to facilitate the use of the library to all its users, acquaint them with all the services offered and the use of the electronic catalog of the library referred to by the OPAC or the on-line public access catalog, the use of electronic resources, multimedia resources.

Specialized Tours: will be scheduled for professionals and librarians to learn the different systems the library is implementing as well as other related matters, which might interest them; like Work Procedure; Human Resources Development programs; Collection Development Policy.

Membership Services: Your membership card gives you the benefit of using the library and all its facilities. You can borrow books, use reading areas, and reserve study rooms, if you qualify.

The Membership Card can also allow you reduced rates to enter Museums, Exhibits and other Cultural Performances.

Specialized Information Services: There are ten specialized information desks designed to serve users and researchers. Other than the three libraries for the Young People, Children, and the Blind or visually impaired.

Researchers Information Services: Each level is designed to serve the generate reader with a collection of material if a certain subject or subjects displayed in the open shelve. Researchers may also reserve study rooms to work in an environment surrounded by books & printed reference sources, with access to selected databases and other electronic resources.

Other Services: Photocopying: users and researchers can purchase pre-paid cards at the General Information Desk to do their own photocopying in the dedicated areas on level B2 and B4. Users & researchers are kindly requested to follow instructed copyright rules and regulations. Printing: users and researchers can also print from their workstations and pay for the print outs, or else buy floppy discs from one of the information desks in the reading areas [25] .

4.5. Development of Library Collection

As the main mission of the library is to be the window of Egypt on the world, and the window of the world on Egypt, and to be a library for the digital age. The library's cataloging and acquisition practices is strongly affected by this mission. Thus, the library collection varies between books available directly to be accessed for browsing and selecting (open stack—VOS), and books limited to the staff of the library or to a limited group of library users (closed stack—VCS). Also, other types of collection are available such as, a digital collection (DC), electronic books (EB), online databases (OD), printed scholarly journals (PSJ), daily newspapers (DN), electronic journals (EJ), theses (TS), specialized library's collection (SLC), and depository library collection (DLC).

As Table 3 shows, the collection of Bibliotheca Alexandrina changed over time. For example, the volumes in open stacks was at its minimum value in 2003 with around 201 thousand book, and reached it maximum value in 2012 with about 288 thousand book, with an average estimated to be 240 thousand book added every year. The same goes for the volumes in closed stacks as its minimum value was in 2003 with around 19 thousand book, and reached its maximum value in 2012 with about 200 thousand book, having about 62 thousand book as an average annual addition. On the other hand, it seems that in the recent years, most of the volumes were added to the library collection as closed stacks as it suddenly jumped from 62 thousand book in 2010 to be around 152 thousand book in 2011, as Figure 7 shows.

In general, all of the other collection types such as the digital collection, electronic books, online databases, printed scholarly journals, daily newspapers, electronic journals, theses, specialized library's collection, and depository library collection has been increased over time but in different ranges.

4.6. Development of Visitors and Membership

The BA receives a numerous number of visitors annually, some of them are Egyptians while others are foreigners, and some of them individual visitors while others are guided tours, beside some of the visitors are VIPs, or even school students. Moreover, its websites receive more than two million hits every day. And it holds an average of around 700 cultural events every year.

Table 3. Collection development in BA, in thousands except for OD and DN.

Year	VOS	VCS	DC	EB	OD	PSJ	DN	EJ	TS	SLC	DLC
2003	201.35	18.86	5.09	15.90	27.00	1.25	72.00	21.45	18.34	42.46	10.66
2004	217.85	20.86	5.59	17.34	32.00	1.04	81.00	23.47	19.45	58.89	11.46
2005	233.10	24.16	6.30	20.00	28.00	1.12	78.00	24.80	20.06	55.87	12.46
2006	228.67	28.57	6.40	30.00	30.00	1.28	85.00	25.00	33.81	60.10	13.21
2007	209.82	37.95	7.00	36.58	31.00	1.27	97.00	35.68	45.98	60.37	18.46
2008	227.79	37.84	6.50	43.69	40.00	4.34	87.00	48.38	54.83	81.55	21.45
2009	248.65	42.38	6.58	59.70	44.00	6.25	88.00	49.40	64.70	90.78	23.99
2010	270.54	61.99	8.40	84.95	41.00	7.81	87.00	50.80	76.65	101.06	25.63
2011	280.25	152.00	8.48	95.48	44.00	7.81	88.00	65.05	87.16	108.98	26.89
2012	287.69	200.25	8.48	111.04	44.00	7.92	88.00	67.19	93.95	112.75	27.05

Source: [26].

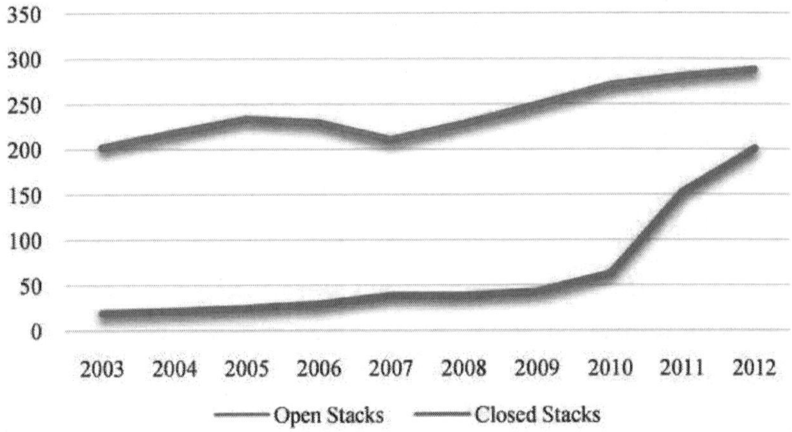

Figure 7. Development of open stack & closed stack collection in BA. Source: [26] .

As for the membership services, which provide a membership card that gives the benefit of using the library and all its facilities. borrow books, use reading areas, and reserve study rooms, if the member qualify. The Membership Card can also reduce rates to enter Museums, Exhibits and other Cultural Performances [25] .

Number of visitors changes overtime, as it was at its minimum value in 2012 with about 107 thousand visitors, while it reached its maximum value in 2009 with about 320 thousand visitors. As for the membership, its minimum value estimated to be around 8 thousand members in 2003, and reached its maximum value in 2011 with about 18 thousand members, as Table 4 shows.

From Figure 8, it was clear the drop that occurred in the BA number of visitors from 227 thousand visitors in 2011 to reach about 107 thousand visitors in 2012, such behavior maybe explained as a result of the political events that happened in 2011 and the revolution which affected BA among other organizations & facilities around the country.

4.7. Staff and Library Management

There are about 2000 staff members in BA, from 30 - 35 as an average, from both genders, and they are distributed between different categories such as senior and junior managers, security and service staff, consultants, and ordinary staff members.

Overall, the staff members tend to increase over time from 1639 members in 2006-originally was about 1200 members in 2003, to reach 2288 members in 2012, as Table 5 shows.

Table 4. Visitors and membership of BA.

Year	Visitors	Membership
2003	231,921	8076
2004	151,573	10,840
2005	196,923	12,845
2006	247,959	12,938
2007	248,667	9835
2008	317,174	11,396
2009	319,500	15,438
2010	310,000	15,898
2011	227,000	17,638
2012	106,800	16,322

Source: [26].

Table 5. BA staff categories.

Year	Sr. manager	Jr. manager	Staff	Security & service staff	Consultants	Total
2006	43	120	832	613	31	1639
2007	45	160	902	618	25	1750
2008	45	156	1076	632	27	1936
2009	47	168	1118	649	33	2015
2010	66	207	1094	823	36	2226
2011	68	250	1270	795	30	2413
2012	48	223	1300	695	22	2288

Source: [26].

5. RESULTS & DISCUSSION

For the first step, results show that the relation between GDP and the allocated budget is linear relation as it is plotted in Figure 9.

Moreover, results show that Pearson's r is 0.96 which proves the existence of a strong relation between the country GDP and the government allocated budget to BA.

As for the second step, which testing the correlation between the allocated budget from one side, and (salaries & wages, general & administration expenses) from the other side. Results summarized in Table 6.

As the shows, there is a strong positive relation between the allocated budget and the salaries & wages as Pearson's r = 0.96, but on the other side, the relation between the allocated budget and the general expenses is insignificant. Which explains how library staff and librarians are very likely to be affected by any budget reallocation, which in turn affected by the economy GDP.

Lastly, results again shows a significant positive relation between the allocated budget and all of the other library indicators except for two of them, which are (the number of visitors, and the daily news), where they showed insignificant relation, as Figure 10 illustrates.

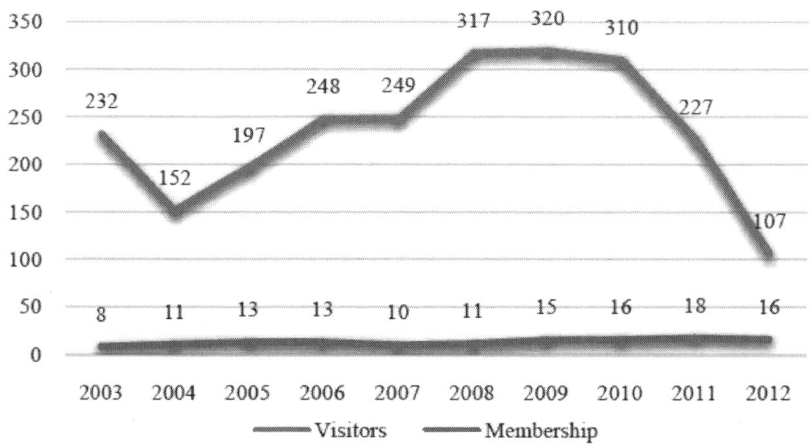

Figure 8. Number of visitors and members of BA. Source: [26] .

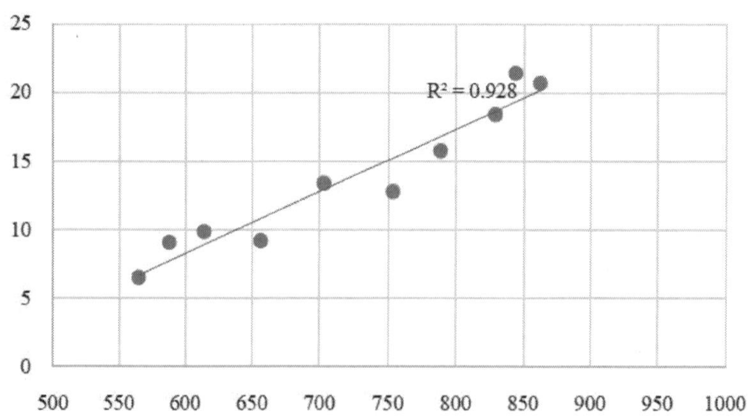

Figure 9. GDP & allocated budget scatter plot.

Table 6. Pearson correlation results.

Pearson correlation	Allocated budget	Salaries & wages	General expenses
Budget	1	0.957**	0.595
Salaries & wages	0.957**	1	0.748*
General expenses	0.595	0.748*	1

**Correlation is significant at the 0.01 level (2-tailed); *Correlation is significant at the 0.05 level (2-tailed).

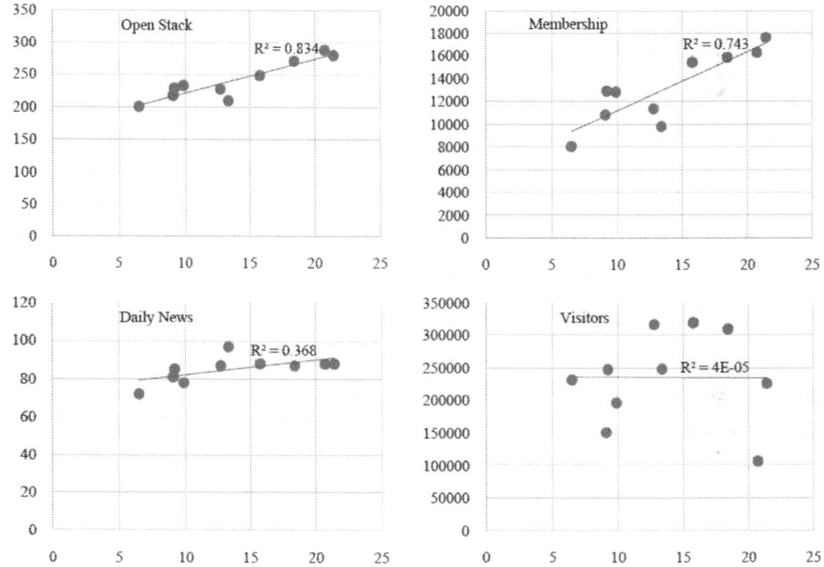

Figure 10. Pearson correlation results for some of the library indicators.

For example, Pearson's r equal 0.914 in the open stack books, which shows a strong relation between the allocated budget and this collection item. On the other hand, Pearson's r is (−0.006) in the visitors' analysis, which is insignificant.

6. CONCLUSIONS

Libraries such as Bibliotheca Alexandrina are a vital component in their local societies. The country economic development on the other hand could impact those components, and affects their ability to deliver sufficient services either directly or indirectly. To investigate this sort of impact, the research extracted data from the annual reports of Bibliotheca Alexandria about various library services, collections, and financial data. As for the data concerning the Egyptian economy, it has been extracted mainly from World Bank data.

Trying to prove such relationship, an analysis has been conducted over the data from 2003 until 2012. Using Pearson product-moment correlation coefficient, the analysis was conducted on two steps. The first one shows the relation between the country GDP and the library allocated budget. Then the second step takes place to investigate the correlation between the allocated budget and various library indicators.

Results show that Pearson's r is 0.96 which proves the existence of a strong relation between the country GDP and the allocated budget. Then, for the second step results show that there is a strong positive relation between the allocated budget on one hand and the salaries & wages on the other hand, which explains how library staff and librarians are very likely to be affected by any budget reallocation, which is in turn affected by the economy GDP.

Except for a number of visitors & the daily news, most of the library indicators show a strong correlation with the allocated budget such as (open stack books, closed stack books, membership, etc.).

The descriptive analysis results also show that the government budget constitutes the bigger part of the library overall revenues, and other parts such as donations & the library own investments and projects constantly decreased over time, which increase the dependence on the government allocated budget even more. Furthermore, if the library wants to achieve its own independency and decrease the government budget reallocation's risks, it should invest more in its own projects and investments.

REFERENCES

1. UNESCO (2016) UNESCO Public Library Manifesto.
2. Kamer, P.M. (2005) The Economic Value of the Northport-East Northport Public Library in Suffolk County. Long Island, New York.
3. Indiana Business Research Center (2007) The Economic Impact of Libraries in Indiana. Indiana Business Research Center, Indiana.
4. Glass, R.H., Clifford, N., Harris, B. and Rose, C. (2000) The Role of Public Libraries in Local Economic Development. The University of Kansas, Policy Research Institute, Kansas.
5. Aabo, S. (2011) The Value of Public Libraries: A Socio-Economic Analysis. Editrice Bibliografica, 169-176.
6. Santos, V.F.D. (2009) Public Libraries and Their Contribution towards Economic Development: A Discussion. Library and Information Science Research Electronic.

7. Martin Prosperity Institute (2013) So Much More: The Economic Impact of the Toronto Public Library on the City of Toronto. Martin Prosperity Institute, Toronto.

8. Barron, D.D., Williams, R.V., Bajjaly, S., Arns, J. and Wilson, S. (2005) The Economic Impact of Public Libraries on South Carolina. The School of Library and Information Science, University of South Carolina, South Carolina.

9. The Urban Institute (2007) Making Cities Stronger: Public Library Contributions to Local Economic Development. Urban Libraries Council, Washington.

10. Steffen, N., Lietzau, Z., Lance, K.C., Rybin, A. and Molliconi, C. (2009) Public Libraries—A Wise Investment: A Return on Investment Study of Colorado Libraries. Library Research Service.

11. Stephen, A.R. (2003) Financial Management of Libraries: Past Trends and Future Prospects. Library Trends, 462-493.

12. Edward, A.W. (1963) Financial Support of Public Libraries. Library Trends, 343-352.

13. Anne, M.-J. (2009) The Impact of Economic Recessions on Libraries: A Past, Present, and Future View From Minnesota. LIBREAS, Library Ideas.

14. Balina, S. (2014) Public Libraries—Facilitators of Information Society and e-Inclusion in Latvia. Procedia—Social and Behavioral Sciences, 109, 412-416.

15. Kate-Riin, K., Andres, K. and Signe, J. (2012) Financing of Estonian Research Libraries: The Happy Days and the

Economic Crisis. Slavic & East European Information Resources, 73-88.

16. Denise, E.A. (2015) Alternative Funding for Public Libraries: Trends, Sources, and the Heated Arguments that Surround It. Influence of Funding on Advances in Librarianship, 31, 115-140.

17. State Information Services (2016) Overview on the Modern History of Egyptian Economy. http://www.sis.gov.eg/En/Templates/Articles/tmpArticles.aspx?CatID=1353

18. Trading Economics (2016) Egypt GDP Growth Rate. http://www.tradingeconomics.com/egypt/gdp-growth

19. World DataBank (2016) World Development Indicators. http://databank.worldbank.org/data/reports.aspx?source=2&country=EGY&series=&period=#

20. Bouri, E.N. (1994) Public Libraries Development Reconsidered: The Case of Egypt. International Information & Library Review, 26, 151-168.

21. Smith, S.E. and Rasol, H.M. (2009) A Geographic Assessment of Public Libraries in Egypt. World Applied Sciences Journal, 6, 861-867.

22. Central Agency for Public Mobilization and Statistics (2016) Public Services Statistics.

23. Desouki, Y. (2007) The Bibliotheca Alexandria and the Preservation of Egypt's Visual Cultural Heritage. New York University, New York.

24. Wikipedia (2015) Bibliotheca Alexandrina.https://en.wikipedia.org/wiki/Bibliotheca_Alexandrina

25. Serageldin, I. (2002) Bibliotheca Alexandrina: The Rebirth of the Library of Alexandria. Bibliotheca Alexandrina, Alexandria.

26. Bibliotheca Alexandrina (2003-2012) Annual Report-Various Versions. Bibliotheca Alexandrina, Alexandria.

27. Bibliotheca Alexandrina (2012) Annual Report. Bibliotheca Alexandrina, Alexandria.

28. Serageldin, I. (2006) Born Digital: The New Bibliotheca Alexandrina. Bibliotheca Alexandrina, Alexandria.

2

CHAPTER

Library3.0 for Public Library

Hubert C. Y. Chan

The Hong Kong Polytechnic University, Hong Kong, China

ABSTRACT

The number of visitors to public library has been dropping in most of the developed cities as a result of digitalization and Internet proliferation. The role of public library has to be converted into a place for discourse, peer collaboration, social learning, and particularly inspiration through and learning from people. Social capital can thus be built up and that the process of building social capital can be part of a community development. Base on the principles of Library3.0 and Connectivism which is considered as the learning theory of this digital age, and the study of "Next" public library, this paper supports that Library3.0 implementation can facilitate this conversion with emphasis on the importance of Personalization in the creation of different space/zones for individual. Library3.0 in reality is introduced at the end.

Keywords: Web3.0, Library3.0, Learning Theory, Connectivism, Public Library

1. INTRODUCTION

The "classic" role of library consists in procuring, keeping up with, making accessible, relaying and archiving academic literature. It is considered as an institution of information supply with a specific information structure that operates as an intermediary between supply and demand on diverse (information) markets (Hanekop and Wittke, 2006) [1] . In the digital age, information can be disseminated through a variety of medias besides hardcopy books or journals. Library infrastructures have been modified to cope with these changes along all these years. Such modifications have increased the library expenses (Kumar and Grover, 2007) [2] . In order to accommodate the needs of users, libraries are required to increase their labor force to attend to acquisitions, promote and manage the e-journals. They must also be tech-competent to handle such responsibilities. The cost of maintaining the e-journal collection is much greater than the costs saved from printing (Montgomery and Beilec, 2000) [3] . As a result, the traditional role of library, librarian, and cost allocation have been changing in this digital age. Besides digitization, the infrastructure investment must be prioritized to achieve an economic effectiveness.

E-lending platforms such as Amazon, Apple and more have offered packages to allow readers to borrow an e-book monthly on their Kindles for free (The Economist, Folding Shelves, 2013) [4] . With such platforms, libraries experience significant competition. Furthermore, modern day technology offers libraries with threats and opportunities. It first began when Web2.0 was released, introducing the library2.0. The characteristics of Web2.0 are rich user experience, user participation, dynamic content, metadata, scalability, openness, freedom and collective intelligence by way of user participation (Best, 2006) [5] while Library2.0 is defined as a subset of library services designed to meet user needs precipitated by the direct and peripheral effects of Web2.0 (Habib, 2006) [6] . The integration of libraries and social media sites provided a platform for discussion and enhanced the online presence of libraries. Users are able to receive updates, ask questions and share content with others conveniently through social media. With that, library is able to connect users through a variety of medias such as Skype, instant messaging, texting, social media and more for discussions. This allows library services to be more accessible as websites operate 24/7. At the same time, interaction between users and the library should be improved because they are able to provide feedback, participate and take part in the community (Vacek, 2007) [7] . Web2.0 was further developed into Web3.0. It combines the semantic web, Web2.0 applications and artificial intelligence, creates the opportunities and possibilities for the use of semantic tagging and annotation for the social web. Librarians have been using Web2.0 for organizational purposes and providing services to users. It has been serving the academic electronic library focused on the SOA (Service-Oriented Architecture), from users' perspectives (Yang et al., 2009) [8] while the application of Web 3.0 characteristics into the core functionalities of library information systems is emerging. The deployment of intelligent agents to form the semantic social web can interpret linguistic expressions of e-library users without ambiguity. In addition, the development of the real world web has enabled mobile phone connection to information service providers. Recently, Bath University in the United Kingdom incorporated QR codes into their library catalogue so that users could save book related information, such as author, title and location onto their smartphones. Users are even able to monitor library traffic through their mobile devices and find out what facilities are available at a specific time. All these allow the evolution of library2.0 to library3.0 which refers to the use of emerging technologies such as the semantic web, cloud computing, mobile devices and established tools like federated search systems, to facilitate the development, organization and sharing of user-generated content through seamless collaboration between users, experts and librarians (Belling et al., 2011) [9] . Apparently, the expansion of the digital world discourages library visits. In recent years, it has been seen that the number of visits to public library has been following a decreasing trend around the world. For instance, the visits made in person to public libraries in Ontario, Canada have dropped by about 7.5% over the course of four years. The number of active cardholders is also decreasing. Over the four years, it has decreased by about 4.4% as shown in Table 1.

Same phenomena are happening in the UK, the annual number of visitors to the library has decreased by 40 million in four years (Morris, 2015) [11] . Not only have visitors to libraries decreased, visitors to library websites have also decreased. Visits to library websites decrease at a rate of 14.4% annually (Murdoch, 2012) [12] . This is the result of digitalization which saves the users a visit to library in person but creates a dilemma. As library has to be functioned as a place for discourse, peer collaboration, social learning, and particularly inspiration through and learning from people is more important than ever before in this knowledge base society (Bilandzic and Foth, 2013) [13] . Can the implementation of Library3.0 achieve this objective?

2. LEARNING THEORIES AND SOCIAL CAPITAL

To answer this research question, a review of Learning Theories is necessary. Learning theory is defined as a framework that helps us think about how and why change occurs (Smith, 2003) [14] . There are four main broad learning theories in the education literature, namely Behaviorism, Cognitivism, Constructivism, and

Table 1. Number of visitors to libraries in ontario. (source: "public library statistics", ontario ministry of tourism, culture and sport, n.d. 05 Aug, 2015) [10]

	2010	2011	2012	2013
Number of Active Cardholders	5,133,356	4,971,127	4,913,281	4,908,914
Library Visits Made in Person	77,719,500	75,058,800	72,529,650	71,928,150

Connectivism. In Behaviorism knowledge is perceived as facts that can be transferred from teacher to student, however, learning is largely unknowable, that is, we can't possibly understand what goes on inside a person. Cognitivism considers the learner as an information processor, opening up the black box of the mind; learning is managed in short term memory, and coded for long-term recall. Whereas Constructivism assumes that learners are not empty vessels to be filled with knowledge. It suggests that learners create knowledge as they try to make meaning of their experiences. Connectivism is considered to be the learning theory of the digital age, "a successor to behaviorism, cognitivism, and constructivism". It is an integration of the principles explored by chaos, network, and complexity and self-organization theories. Principles of Connectivism (Siemens, 2004) [15] are:

- Learning and knowledge rest in diversity of opinions.
- Learning is a process of connecting specialized nodes or information sources.
- Learning may reside in non-human appliances.
- Capacity to know more is more critical than what is currently known.

- Nurturing and maintaining connections are needed to facilitate continual learning.
- Ability to see connections between fields, ideas, and concepts is a core skill.
- Currency (accurate, up-to-date knowledge) is the intent of all connectivist learning activities.
- Decision-making itself is a learning process. Choosing what to learn and the meaning of incoming information is seen through the lens of a shifting reality.

The principles emphasize on connections with information sources, different fields, and ideas. This connection can be made through social networks which are built on the premise of collaboration and sharing. They are ideal for connectivity particularly in external learning situations that are not always available, or feasible, in face-to-face classrooms (Mallon, 2013) [16] . It has been found that there is a strong positive effect of social capital on student satisfaction toward online learning and on virtual group learning outcomes (Lu et al., 2013) [17] . The concept of social capital is defined as an intangible asset produced by social relationships as a result of social networking (Coleman, 1994) [18] . It can also be defined as a feature of social organization including trust, norms, or networks that improve the efficiency of society by facilitating coordinated action and reducing transaction costs. There are two forms of social capital, namely bonding social capital, and bridging social capital. The former refers the thick confidence (particularized trust) between members of tight and highly-integrated groups while the latter implies the thin confidence (generalized trust) across primary belongings. It has been indicated that frequent library users have a higher social capital than those who use the library less often. Frequent library users are more apt to engage in civic life by voting, and are more trusting. As a result, social benefits gained from library use spill over into the community and add to the general quality of community life because the use of the library is a deliberate, if not always conscious, strategy to engage with fellow human beings (Johnson and Griffis, 2009) [19] . Learning experience is much more fruitful through conversation and interaction, on sharing, creation and participation as embedded in meaningful activities such as games or workflows (Downes, 2006) [20] . Furthermore, Hussain (2012) [21] studied the evolution of e-Learning riding of that of Web "point oh", and argued that a call for a new learning theory is not justified since most of the technologies that are to be a part of e-Learning are addressed by the principles of Connectivism. The same argument is applicable to Library3.0 if its principles can align with those of Connectivism.

3. LIBRARY3.0

School library is a private realm where dominating relational form is between students of the same school. Whereas, public library is a public realm, most of the visitors are strangers to each other, and the library's community activities constitute it as a parochial realm. Both types have to change their scope of services and tasks as a result of digitizing and digital swift. The evolution is not only technology driven due to that of Web and e-Learning "point oh", but also market driven as the core skills include creativity, interdisciplinary thinking, problem solving, and the ability to collaborate with others cannot be learned easily from books, but rather through learning-by-doing and social interaction in today's knowledge base economy. Not to mention most users work individually or within pre-organized groups who do not usually make new connections with co-present, unacquainted users (Bilandzic and Foth, 2013) [13] , the attraction of traditional library learning arena is diminishing. Even with the adaption of Web3.0, electronic library can provides a new way for end users to collect online contents concerning comments, articles, Bolgs, photos, and videos in one place through parsing the linguistic expressions provided without the intentionally invoke specific application programs (Chu and Yang, 2012) [22] , it cannot fully convert a traditional library into an information center, a meeting place, and a technology hub. Library should embrace Web3.0 and go beyond. The facilities of library thus have to be changed as well as the role of librarian. It seems that the role of librarian has been weakened in the Library2.0 environment, they must be aware of the use of artificial intelligence management tools and semantic web in Library (Ahmed, 2015) [23] . The new role of librarian is collecting & providing information and offering instruction in research techniques and information use rather than by physical presence as somewhere people go (Saw and Todd, 2007) [24] . They are not only the prominent apomediaries guiding library users on how best to locate, access and use credible information in myriad formats from diverse sources, at the point of need (Kwanya et al., 2013) [25] , but also facilitators of connecting users. To convert a library into a personalizable, intelligent, sensitive and living institution with seamless engagement of library users, librarians and subject experts on a federated network of information pathways, Kwanya et al. (2013) [25] , posited five principles of Library3.0, namely the library is intelligent, organized, federated networked, apomediated, and personalized. The meaning and attributes of each principle are listed in Table 2 below.

Table 2. Summary of library3.0/3.1 principles, attributes, and applied context-awareness (adapted from kwanya et al., 2013 [25] ; noh, 2013 [26] ; siemens, 2004 [15] ; and hussain, 2012 [21]).

Library3.0 principles	Attributes of Library3.0	Applied context-awareness	Principles of Connectivism
The library is intelligent: Self-renewing, flexible, functional, integrated, efficient, resilient, autonomous and sensitive (adaptive)	Self-renewing, flexible, functional, integrated, efficient, resilient, autonomous and sensitive (adaptive) via selective intelligence (wisdom of the expert), ontology, smart (intuitive) searching in natural language, artificial intelligence	Space on/off: Controlling the space according to user access. Turn off the light & other available technology once the user entering the space. Lobby: Direct users to their desired space using location based technology	Learning may reside in non-human appliances.
The library is organized: Turn the unorganized web of information into a systematic and usable body of knowledge by exhaustively describing and linking every piece of data to enable ease of access	Turning the unorganized web of information into a systematic and usable body of knowledge by exhaustively describing and linking every piece of data to enable ease of access via semantic Web/Search		Ability to see connections between fields, ideas, and concepts is a core skill. Nurturing and maintaining connections is needed to facilitate continual learning
The library is a federated network of information pathways: Unifies disparate information channels, formats and environments to ensure availability, accessibility, search ability and usability of credible information	Supports expressiveness and interoperability to create synergies between hitherto disparate information resources and systems via network of data (meaning), all-visible, all-accessible Web, and Cloud computing		Learning is a process of connecting specialized nodes or information sources. Currency of knowledge is important.
The library is apomediated: Stand by the users and guide them to high quality information and services	Stand by the users and guide them to high quality information and services without overbearing on the users via social media, peers, & librarians.	Reference & Safety services: Riding on location based technology, if the user spends over certain duration at specific location looking for specific subjects, recommendations will be sent to him/her accordingly. If the user doesn't move over a certain time, alert will be sent to library personnel.	
The library is "my library": Personalization similar to what various service pro-viders such as search en-gines and the media were offering their clients on various Internet platforms	Personalization of library services via the design, management and delivery of content based on known, observed and predictive information. The opportunity to interact with other library users and to form or join groups to enrich their learning experience	Access service: The system detects the user identity by reading the RFID card, & then pushes the recommendations according to his/her borrowing history.	Learning and knowledge rests in diversity of opinions. Capacity to know more is more critical than what is currently known

To develop an intelligent system which can search and provide the services that users require in their current situation by analyzing and identifying the available contextual information such as the user's current location, time, people and devices in the vicinity, and the user's behavior and inputted data, Context-Awareness must be applied (Noh, 2013) [26] . He posited the use of context awareness as a core concept of the next generation digital libraries. Context is divided into "user context" and "environment context". The former represents the wide range of information described by the user and describing the user while the latter is the contextual information relating to the user's current user's current state, such as location or time zone, which cannot be described within the user's context (Hartmann, 2010) [27] . In short, Context-aware computing provides service to meet the needs of the users by recognizing certain situations, such as the user's location or environment. The author expounded the applications into five proposals: Applied context-awareness access service, Space on/off according to the user's access, Applied context-awareness lobby, Applied context-awareness reference service, and Applied context-awareness safety service. However, Noh (2013) [26] realized that Context-awareness has not been fully implemented in library yet, and thus the performance required further evaluation in following studies. The Principles of Library3.0, Context Awareness, and Connectivism are summarized in Table 2. It has been shown that the principles of Libray3.0, Context-Awareness, and Connectivism are aligned with each other.

Although the total number of library visitors has been declining in recent years as mentioned before, research has shown that the number of children borrowing books from libraries is at its highest in history. There are children zone where children can read, talk, and play on the floor. Studies have also shown that students prefer actual books to virtual books. Libraries remain to be the source of information, culture and wisdom (Gaiman, 2013) [28] . Another group of frequent library users are the elderly who may not be familiar with Web searching. They are healthy, retired, prefer to read newspaper and enjoy the air-con in the summer time but seldom make friends with other users. Regret that the elderly is committing more crimes than those aged 14 to

19 in recent Japan. This is because some elderly would be willing to commit crimes just to find companions and care in prison (The Economist, Victims' justice, 2015) [29] . Elderly loneliness is a common problem in this aging society. Library is a place which can mitigate this problem particularly for those who do not stay in elderly home by providing some group activities other than reading materials. For instance, just the like children zone, libraries can set up themed corners such as the smart senior room to group individuals with different interests and interactions; it creates companion opportunity for them. The smart senior room is equipped with rehabilitation computer programs in which seniors will be able to select music, digital instructor, 3D animation, so that participants can go through the program at their own pace. At the same time, interactive games involving the multi-touch display for multiple players; different activities, and events can be organized for interaction and fun. As such, in order to change libraries for books/medias in industrial society to libraries for man/human in networked/ knowledge society, it should provide different spaces/zones to cater the need for individual, quite space for reading, noisy space for group meeting, learning, meeting, discussion, and performance for mixed or different age groups, partnership with creative alliance to organize different types of function, and integrate with citizen services to attract users.

Schulz (2014) [30] realized the above societal demand. He posited a new model for the "Next" public library, providing "Experience, Discover, Participate, and Create" through digitization and individualization: Experience through inspiration space, Discover through learning space, Participate through meeting space, and Create through performance space. Inspiration space should open up to the irrational, emotional and chaotic by communicating a diversion of aesthetic experiences through storytelling or other artistic expressions within all kinds of media, cultural forms and genres. Learning space should provide experience and exploration of the world and thus strengthen their competences and possibilities through free and easy access to information and knowledge, music, play and many other activities. Meeting space is an open arena for meeting other people with different interests and values and encounter opinions that challenge you through discussions and debate in non-committal, random encounters through lounge décor with newspapers and café atmosphere as well as more organized meetings. Performance space can provide user interaction with others for inspiration and creation of new creative expressions in the encounter with art and culture through interactive games and writing-, sound- and video-workshops. Library services can then be tailored for personal needs and circumstances as well as giving users the opportunity to interact with other library users and to form or join groups to enrich their library experience (Storey, 2004) [31] . This echoed the findings of Aabø and Audunson (2012) [32] who explored how the libraries function as meeting places for across cultural, ethnic, generational, and social lines in a complex multicultural and digitized society. They found that a high proportion of the users constitute the library as a place related to work and education. Variety of users making the library stands out as a conglomerate of the community. Some people working alone use the library as their work space while some people in between jobs use it for job- searching. Some people periodically use the public library as an inspiring place to work. It is an extension of the school, home or work place. Social capital can thus be built up and that the process of building social capital can be part of a community development process (Kilpatrick et al., 2003) [33] . In fact, all these aligned with the principle of "Personalization" in Library3.0 (Kwanya et al., 2013) [25] . Schulz (2014) [30] further illustrated the integration of citizen services such as issuing passport, driving license, and health security card with library can bring citizen in contact with library. Some cities like Oslo and Helsinki built concert hall and museum side by side with library so as to create a new icon in the city. In fact, a number of new public libraries all over the world had adapted the "Space" concept in one way or the others in their designs. To name a few: Vennesla Library in Norway is located in the city centre provides café, open meeting places, and a small scene, the building combines a library and house of culture. Black Diamond Library in Denmark includes a 600-seat auditorium, the Queen's Hall, used for concerts, literary events, theatrical performances and conferences. Surry Hills Library in Australia specializes in fashion and design titles, as well as having a significant lesbian, gay, bisexual and transgender collection. Nam June Paik Library in Korea aims to promote "non-linear and random access to information, and its production beyond the prescribed consumption." The user of information becomes the producer, and the static contents of the library turns dynamic through spontaneous expression and juxtaposition of ideas. The interior design of Biblioteca España Library in Colombia could "decontextualize the individual from the poverty that is experienced in the outside." The Maranello Library in Italy is intended to provide its visitors an introspective experience and visual connection with nature. The library of Muyinga in Burundi was built with the local know-how and traditions of Muyinga in mind. A "hallway porch" along a circulation space is often a part of the Burundian traditional housing as it provides a shelter from heavy rains and harsh sun.

4. LIBRARY DEVELOPMENT IN REALITY

A full implementation of Library3.0 is still remote. In fact, there is no consensus on the actual definition of Library3.0. Build on the Library3.0 generation evolution graphical representation by Noh (2012) [34] , this paper is trying to elaborate the technology available in the market for library in Figure 1, followed by a more detail explanations in Table 3.

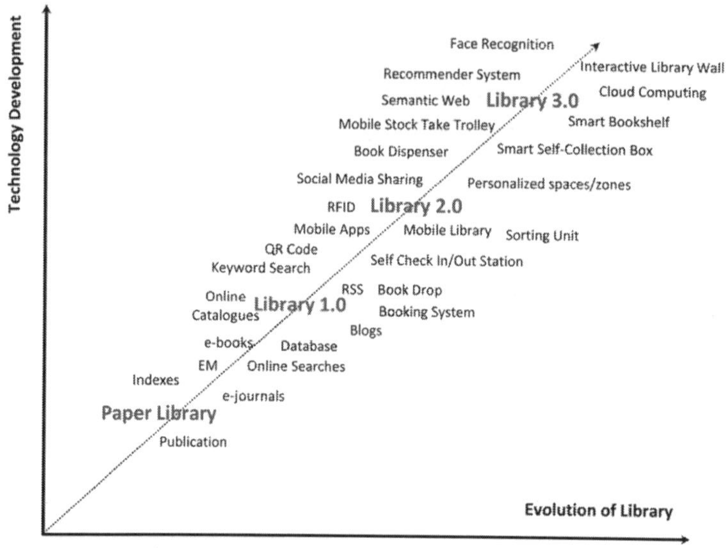

Figure 1. Evolution of library technology (adapted from Noh (2012) [34] with additional updates).

4.1. The Library Is Intelligent and Organized

The attributes of "Intelligent" of the principle of Library3.0 are Self-renewing, flexible, functional, integrated, efficient, resilient, autonomous and sensitive (adaptive) via selective intelligence (wisdom of the expert), ontology, smart (intuitive) searching in natural language, and artificial intelligence while those of "Organized" are Turning the unorganized web of information into a systematic and usable body of knowledge by exhaustively describing and linking every piece of data to enable ease of access via semantic via Web/Search (Kwanya et al., 2013 [25] ; Noh, 2013 [26]). The two principles are interrelated with each other in reality.

Table 3. Products for library3.0.

Intelligent & organized		
Product	**Features**	**Photo**
Library Automation	• Detects room occupancy • Automatic scene setting for different functions • Measures brightness & adjust lighting or curtain position • Detects temperature in room and adjusts to attain energy saving • Intercom system for communication with librarian • Integrated with room reservation & door access system	
RFID Applications	• RFID tags inside each book • Gate keeps track of RFID materials passing for security check • Counts the number of visitors	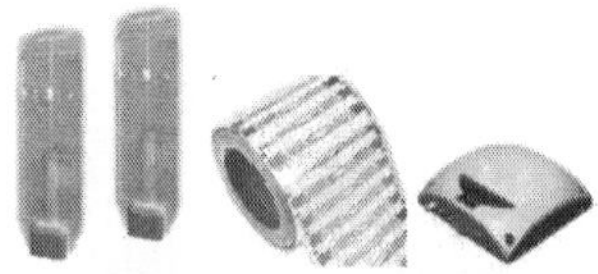
Smart Bookshelf	• Automatically detects on-shelve books • Updates system records for stock take • On-shelve books searching	
Smart Self-Collection Box	• Automatically distributes reserved books into dedicated smart shelf holes • Real time tracking and monitoring • Generates alerts in the case something has been misplaced	
Mobile Stock Take Trolley	• Automatically detects books on the trolley for uploading to shelf • Provides stock functions with a handheld device	
Book Drop and Sorting Unit	• 24-hour book drop • Sorts books into their designated categories • Improves library efficiency and book location accuracy	

Web3.0 enables automatic clustering based on similarity and relationships in the document content for easy searching. Some libraries are using Resource Description Framework (RDF), Ontology, Resource Description and Access (RDA) to organize and describe their collections. In the hardware side, the development of internet of technology (IoT) has impacted library models and its operation. The IoT library consists of information object, independent network and intelligent application. Information object refers to library literature resources. Its items include hardware facilities and security access control. The independent network denotes network systems with self-management abilities that have exhibit self-configuration, self-healing, self-optimization and self- protection capabilities. Libraries will offer learning experiences through self-serving and automation, efficient operation, user generated or personalized content, mobility and access 24/7. People of all ages will be able to experience and explore the world through convenient access to information and knowledge. Intelligent terminals help readers to use the library resources more efficiently, which therefore improves the resource utilization rates of libraries. It includes intelligent navigation retrieval terminal, security access control system, borrowing system and more (Wang and Zhao, 2015) [35] .

To start-off with RFID enabled facility including tag inside each book, self-check in/out kiosks, security gate, and RFID enabled book-drop for return, the library becomes self-service. More advanced kiosk is equipped with a detector that allows the station to adjust its height according to the user's height.

A 24-hour book drop along with a sorting unit improves the efficiency of the operation process. The device is able to efficiently and accurately sort the returned materials according to pre-defined categories. With the book drop and sorting unit, the workload of librarians is significantly reduced.

To make libraries even more efficient, smart bookshelves can be used. The smart bookshelf detects the on-shelf books automatically and updates the library management system database. The book list will then be generated on screen upon user enquiry at the input terminal at the side. The bookshelf also detects the book sequence, which allows for easy book searching. In addition, the mobile stock-take trolley can automatically detect the books on the trolley and informs the librarian of their positions on the shelf. The stock take trolley provides a book searching function and can identify what books are missing from the shelf after the stock-take with a handheld device. More advanced system such as Automated stored and retrieval system (ASRS) making use of robotic device is also available.

The booking system will then allow users to book library facilities such as function rooms and multimedia equipment. It will have customized ready scene settings and door access based on booking arrangements and more. Libraries can apply space on/of according to the user's access to enhance library automation functions. Within the function rooms, motion sensors will be installed to detect the flow of people in the area. Through identifying motion in the proximity, lights and air conditioners can be controlled automatically. When a user accesses the area, the lights and air conditioner will switch on. A light (lux) sensor will then adjust the brightness and temperature of the room accordingly depending on the traffic flow. The concept of "green energy" is not only practical but also educational to library users.

4.2. The Library Is a Federated Network of Information Pathways

The attributes of "Federated network of information pathways" of the principle of Library3.0 are supporting expressiveness and interoperability to create synergies between hitherto disparate information resources and systems via network of data (meaning), all-visible, all-accessible Web, and Cloud computing (Kwanya et al., 2013 [25] ; Noh, 2013 [26]). Information-discovery system has been adopted by some universities to provide access to a large and diverse landscape of materials across different locations. It is flexible and scalable to support collaborative services, such as inter-library loan services and collaborative online reference in a union framework (Minami, 2008) [36] . Besides digital media, physical books in different locations can be delivered to users at their designated locations where the book dispenser is installed. One dispenser can store up around 200 - 400 books. The upload and re-collection of books at the dispensers can be done during office hours while the users can borrow and return books after office hours. The reader (RFID/EM) at the input slot identifies the book and put it in an empty slot with a robotic arm. It retrieves the reserved book upon the user presentation of smart card. In addition, users are able to reserve books in library and ask for delivery to the dispenser at the built-in touch screen or mobile App. Due to the fact that users must use their library cards or smart cards to borrow books, the system will be able to identify their reading preferences and recommend books personalized for the users. Thus, providing applied context-awareness reference services to users.

4.3. The Library Is Apomediated

The attributes of "Federated network of information pathways" of the principle of Library3.0 are standing by the users and guide them to high quality information and services without overbearing on the users via social media, peers, & librarians (Kwanya et al., 2013 [25] ; Noh, 2013 [26]). Library3.0 glorifies disintermediation and self-service. With an Apps installed in the user portable devices and the Real Time Location System (RTSL) engine at the server or Cloud, his/her location inside the library can be monitored in real time, if the user spends over certain duration at specific location looking for specific subjects, recommendations will be sent to him/her accordingly. The user can then be guided to the appropriate location on a digital map shown on the smart device. Staff contacts and other relevant information can also be accessed at the mobile platform. More detail information about library resources, and reference services can be provided through QR codes on the catalogued items. In addition, to ensure the safety of the library and users, CCTVs will be used to monitor any anomaly such as unattended bags or unexpected behavior of users.

4.4. The Library Is My Library

The attributes of "My Library" of the principle of Library3.0 is personalization via the design, management and delivery of content based on known, observed and predictive information (Kwanya et al., 2013 [25]). The role of people evolved from passive consumers of information to that of active contributors because of some aspects of the Web2.0. Through the use of radio frequency identification (RFID) technology, which is a contactless identification technology, libraries will be able to identify users (Wang and Zhao, 2015) [35] . Besides RFID smart card, face recognition can also identify the user at the entrance. The recommender system can then recommend books personalized for each user based on the collected information. This generally involves the application of Machine Learning (ML) techniques, whose goal is learning to categorize new information items based on previously seen information (Lops et al., 2011) [37] . The two common techniques are content-based recommendation system and collaborative filtering. The former tries to recommend items similar to those a given user has liked in the past while the latter is based on the assumption that like minded people who have preferred on past items are likely to agree on future items. A hybrid system combining two methods is also available. Chatbots can even interact with users through intelligent conversation. The recommender system can then use the map presentation agent where the individual's learning locations are displayed on an electronic map guiding them the location of the books, and thus encourage readers to read more and enhance user experience (Chen, 2013) [38] .

Library should provide opportunity to interact with other library users and to form or join groups to enrich their learning experience. It can offer mobile apps with group sharing functions, smart panels for meetings or discussions, and an intelligent function room for sharing to promote social interactions. Sharing and communication will be encouraged if different sessions designed for different age and interest groups can be provided. Library facilities rooms can facilitate group sharing or gathering activities, interest groups for activities like cooking, sports, gardening or music can be formed and sharing sessions can be organized. On the other hand, function rooms can be used to enjoy CDs or DVDs borrowed from the library. They can also be used for academic, business and sharing discussions. Online chat rooms can also be set up for library users to discuss books, movies, interests and more. In addition, different events such as story telling sessions, themed lectures, or themed book, CD recommendation sessions can be hosted. Interactive Multi-touch panels or wall are available for arousing user interests in searching and sharing information.

5. CONCLUSION

The adaption of Web3.0 provides convenience to library user on one hand but discourages library visit on the other. Library3.0 must embrace Web3.0 and go beyond. The role of public library has to be converted into an information center, a meeting place, and a technology hub through which social capital is built-up and spill over into the community adding to the general quality of community life. This study compares that the principles of Library3.0 (Kwanya et al., 2013) [25] and those of Connectivism (Siemens, 2004) [15] ; literature review of the "Next" library, concludes that Library3.0 implementation can achieve this conversion with emphasis on "Personalization" in building communities for different age groups.

6. LIMITATIONS AND FUTURE RESEARCH

The proposal of this study is mainly based on literature review. It should be further verified by interview with library users particularly the elderly. Inclusion of citizen services and integration with other public utilities such as museum, opera, and exhibition hall with library is part the urban planning. This should be further explored at the municipal level.

REFERENCES

1. Hanekop, H. and Wittke, V. (2006) Das wissenschaftliche Journal und seine m&oumlglichen Alternativen: Ver&aumlnderungen der Wissenschaftskommunikation durch das Internet. In: Hagenhoff, S. (Hrsg.), Internet&oumlkonomie in der Medienbranche, Universit&aumltsverlag G&oumlttingen.

2. Kumar, S. and Grover, V.K. (2007) Electronic Journals: Impact on Scholarly Communication, User and Library. Library Herald, 45, 325-336.

3. Montgomery, C.H. and Beilec, J.A. (2000) The Economic Impact of an Electronic Journal Collection on an Academic Library, Research and Advanced Technology for Digital Libraries. Springer, Berlin Heidelberg, 413-417.

4. Folding Shelves (2013) The Economist. http://www.economist.com/news/international/21573966-e-books-mean-plot-twist-public-libraries-and-publishers-folding-shelves

5. Best, D. (2006) Web 2.0: Next Big Thing or Next Big Internet Bubble? Technische Universiteit Eindhoven.

6. Habib, M.C. (2006) Toward Academic Library 2.0: Development and Application of a Library 2.0 Methodology. Unpublished Master's Dissertation, University of North Carolina at Chapel Hill.

7. Vacek, R. (2007) The Impact of Web 2.0 on Library Websites. Slideshare. http://www.slideshare.net/vacekrae/the-impact-of-web-20-on-library-websites

8. Yang, X.Y., Wei, Q.Y. and Peng, X.D. (2009) System Architecture of Library 2.0. The Electronic Library, 27, 283-291. http://dx.doi.org/10.1108/02640470910947629

9. Belling, A., Rhodes, A., Smith, J., Thomson, S. and Thorn, B. (2011) Exploring Library 3.0 and beyond. http://www.libraries.vic.gov.au/downloads/20102011_Shared_Leadership_Program_Presentation_Day_/exploring_library_3.pdf

10. Public Library Statistics (2015) Ontario Ministry of Travel, Culture and Sports. http://www.mtc.gov. on.ca/ en/libraries/statistics.shtml

11. Morris, N. (2015) Number of Library Visitors Falls by 40 Million in Four Years as Austerity Measures Force Closures. The Independent.

12. Murdoch, J.B. (2012) UK Library Stats: Falls in Numbers of Facilities, Visits and Books Issued. The Guardian.

13. Bilandzic, M. and Foth, M. (2013) Libraries as Coworking Spaces: Understanding User Motivations and Perceived Barriers to Social Learning. Library Hi Tech, 31, 254-273.

14. 14. Smith, M.K. (2003) Learning Theory. Retrieved from The Encyclopedia of Informal Education.

15. Siemens, G. (2004) Connectivism: A Learning Theory for the Digital Age. International Journal of Instructional Technology & Distance Learning, 2.

16. Mallon, M.N. (2013) Extending the Learning Process: Using the Theory of Connectivism to Inspire Student Collaboration. Kansas Library Association College and University Libraries Section Proceedings, 3, 18-27.

17. Lu, J., Yang, J. and Yu, C.-S. (2013) Is Social Capital Effective for Online Learning? Information & Management, 50, 507-522.

18. Coleman, J.S. (1994) Foundations of Social Theory. Harvard University Press, Cambridge, MA.

19. Johnson, C.A. and Griffis, M.R. (2009) A Place Where Everybody Knows Your Name? Investigating Relationship

between Public Libraries and Social Capital. Canadian Journal of Information and Library Science, 33, 159-191.

20. Downes, S. (2006) Learning Networks and Connective Knowledge. Collective Intelligence and Elearning, 20, 1-26.

21. Hussain, F. (2012) E-Learning 3.0 = E-Learning 2.0 + Web 3.0? International Conference on Cognition and Exploratory Learning in Digital Age (CELDA 2012), Madrid, 19-21 October 2012, 11-18

22. Chu, H.C. and Yang, S.W. (2012) Innovative Semantic Web Services for Next Generation Academic Electronic Library via Web 3.0 via Distributed Artificial Intelligence. In: Pan, J.-S., Chen, S.-M. and Nguyen, N.T., Eds., Intelligent Information and Database Systems, Springer, Berlin, 118-124.

23. Ahmed, W. (2015) Third Generation of the Web: Libraries, Librarians and Web 3.0. Library Hi Tech News, 32, 6-8.

24. Saw, G. and Todd, H. (2007) Library 3.0: Where Art Our Skills? World Library and Information Congress, 73rd IFLA General Conference and Council, Durban, 19-23 August 2007.

25. Kwanya, T., Stilwell, C. and Underwood, P.G. (2013) Intelligent Libraries and Apomediators: Distinguishing between Library 3.0 and Library 2.0. Journal of Librarianship and Information Science, 45, 187-197.

26. Noh, Y.H. (2013) A Study on Next-Generation Digital Library Using Context-Awareness Technology. Library Hi Tech, 31, 236-253.

27. Hartmann, M. (2010) Context-Aware Intelligent User Interfaces for Supporting System Use. Dissertation zur Erlangung des akademischen Grades. Dr.-Ing.

28. Gaiman, N. (2013) Neil Gaiman: Why Our Future Depends on Libraries, Reading and Daydreaming. The Guardian.

29. Victims' Justice (2015) The Economist.

30. Schulz, K. (2014) Library Transformation: Sketching the Future Library Transitioning to a Space for Citizens and Community Connections.

31. Storey, T. (2004) Libraries Look to Balance Technology, Cost and Usefulness. OCLC Newsletter.

32. Aabø, S. and Audunson, R. (2012) Use of Library Space and the Library as Place. Library & Information Science Research, 34, 138-149.

33. Kilpatrick, S., Field, J. and Falk, I. (2003) Social Capital: An analytical Tool for Exploring Lifelong Learning and Community Development. British Educational Research Journal, 29, 417-432.

34. Noh, Y.H. (2012) A Study on Library 3.0 Concept and Its Service Model. Journal of the Korean Society for Information Management, 27, 283-307.

35. Wang, T.J. and Zhao, P.J. (2015) Research and Application of Internet of Things in Intelligent Library. 1st International Conference on Information Sciences, Machinery, Materials and Energy, Chongqing, 11-13 April 2015, Atlantis Press, 684-687.

36. Minami, T. (2008) Library Services as Multi Agent System. In: Nguyen, N.T., Jo, G.S., Howlett, R.J. and Jain, L.C., Eds., Agent and Multi-Systems: Technologies and Applications, Springer, Berlin, 222-231.

37. Lops, P., de Gemmis, M. and Semeraro, G. (2011) Content-Based Recommender Systems: State of the Art and Trends. In: Ricci, F., Rokach, L., Shapira, B. and Kantor, P.B., Eds., Recommender Systems Handbook, Springer US, New York, 73-105.

38. Chen, C.M. (2013) An Intelligent Mobile Location-Aware Book Recommendation System That Enhances Problem-Based Learning in Libraries. Interactive Learning Environments, 21, 469-495.

3

CHAPTER

Conceptual Principles of the Planning of Modern Public Libraries

K. Jakovlevas-Mateckis, L. Kostinaitė, J. Pupelienė

ABSTRACT

Over the last decades scientific progress has inspired not only major technological but most of the social and cultural changes as well. The information society of new quality has been developed. It could be characterised by the rapid growth of the amount and variety of information, development of the conception of information as a commodity, formation of knowledge and information industry, development of the information and communication technology, the impact of information and its technologies on all spheres of life including the nature of learning and work as well as on the style of everyday life.

INTRODUCTION

Over the last decades scientific progress has inspired not only major technological but most of the social and cultural changes as well. The information society of new quality has been developed. It could be characterised by the rapid growth of the amount and variety of information, development of the conception of information as a commodity, formation of knowledge and information industry, development of the information and communication technology, the impact of information and its technologies on all spheres of life including the nature of learning and work as well as on the style of everyday life.

Today the character of activities and architectural solutions of traditional libraries do not answer the needs of modern information society and particularly its needs on information provision. There is a great lack of modern libraries in Lithuania, the state that fourteen years ago has restored its independence. After some years of deep business depression and crisis following the change from a totalitarian regime into a market economy there can be mentioned some positive moves. The last years of significant rise of state economy has allowed development of the policy of Library Renaissance in Lithuania.

In 2002 the Lithuanian Ministry of Culture has approved the "Library renovation and modernization plan for 2003-2013", which includes the building of 4 new and wide-ranging reconstruction of 5 public library buildings, as well as smaller reconstruction and renovation of several public libraries in the state. Lithuanian librarians and architects have started to discuss what the modern library building should be. Therefore has been organized at the Kaunas University of Technology(KTU) a group of architects and librarians led by Vilnius Gediminas Technical

University (VGTU) professor of architecture, K.Jakovlevas-Mateckis. The co-authors of this article: architect and librarian Mrs.Lina Kostinaitė, and library director of the Klaipėda University (KU), Mrs.Janina Pupelienė can be mentioned as the most active members of this group.

The group analysed the development of Lithuanian libraries and did sociological research on main factors influencing library architecture solutions to develop conceptual principles of modern library buildings. On the basis of the developed conceptual principles a pilot project of a modern regional public library has been made. For wider discussions on these developments the conference at the Kaunas Technology University has been organized on February of 2003 (Šiuolaikinės, 2003). For the first time in Lithuania leading architects and librarians met and discussed together. The participation of Ms. Ewa Kobierska-Maciuscko, director of the Warszawa University Library, member of the Expert Committee of the LIBER Architecture Group, was very important for this conference. She presented the LIBER organization and its activity on library architecture to Lithuanian architects and librarians, evaluated Ms. Kostinaite's pilot project and research developments of the above-mentioned group.

The main purpose of this article is to analyse some aspects of modern public library performance and its architecture, to highlight the main factors influencing its architecture and to present some conceptual suggestions for the planning of these types of libraries.

SOCIOLOGICAL RESEARCH

With the purpose of objective evaluation of the correspondence of current architectural solutions of Lithuanian libraries to the main needs of its users and staff, to identify what kind of library is needed for the current user, the sociological research has been made by the Department of Architecture at the Kaunas University of Technology (Kostinaitė &. Jakovlevas-Mateckis, 2002). Three regional libraries -Lithuanian National Library (LNB), Kaunas County Public Library (KVB) andKlaipeda County Public Library (KLAVB) - have been chosen as the basis for the research. These libraries are situated in a newly developed or developing building. For a long time the first two libraries have been the only libraries in Lithuania with the buildings specially built for library purposes.

There have been presented four blocks of questions to the respondents. These blocks are:

- The evaluation of the library's surrounding.
- The evaluation of the library's interior.
- The evaluation of the library's technologies and functions.
- What the modern library has to be

From these blocks 28 questions have been addressed to the users and 38 to library staff. The analysis has been made from the total of 2500 answered questionnaires. The biggest part (2400) of the answers was given by the library users and only 100 by the staff of these three evaluated libraries.

In general, all the answers are rather pessimistic. Only 18% of users are satisfied with the library's out-door surroundings. 75%

of respondents have mentioned the lack of parking places near the libraries. Libraries' interior satisfies only 37% of users and 39% of staff. Only 34% of library staff evaluated their working places as comfortable and well designed. In all evaluated libraries the lack of modern library equipment and technology was stressed. 59% of respondents evaluated the library building as not corresponding to needs of modern library technology. The lack of places to relax has been mentioned in 59% of respondents' answers. Together with an evaluation of the existing buildings, all respondents expressed their wishes concerning a new public library. Summarizing their answers, a new type of library should be fully equipped with modern information technology, while the library building should be designed to the cultural and social activity of the library users: for language, computer, and business courses; amateur art and social communication of the library users.

We can state that today's library is experiencing a period of changes. In order to survive in such constantly changing environment, every library has to change itself according to the needs of modern information society. That's why the library's environment has to be developed systematically, following the principles of urban planning, architecture, engineering, design and library sciences (Jakovlevas-Mateckis, 2000). Nowadays we understand the importance of the development of an attractive public library by all means: architectural, functional, services provided etc. It has to correspond not only to informational, communicational but also to cultural, leisure, aesthetical and other community needs, to provide not only traditional library services but to organise non-traditional - cultural and social - activity. That's why we have to develop today a new multifunctional type of library according to a flexible model where one third of the library space should be assigned for the additional social and cultural activity.

Main Factors Influencing Architectural Solutions Of Library Planning

The process of planning library buildings revolves around the intersection of architecture and library sciences, each of them with its theoretical and philosophical level. Library architecture combines some factors of these spheres which interaction gives these buildings certain singularity, very typical for this kind of buildings. That's why we can name these factors as integrated. Integrated library factors - library technology, library functions and information media (Kostinaitė, 2001) - determine usually the planned spatial solutions of the library, which are in particular the integrated architectural factors. Otherwise, architectural solutions determine the quality of accomplishment of library functions, the rationality of the technological scheme, the character of deployment of information media etc. So we can state, that interaction of these two types of integrated factors is mutual. This integrated sphere cannot be attached to library science, because the planned spatial solutions of buildings belong to the sphere of competence of architecture. Otherwise, analysis of directly influential factors (library technology, library functions, library media) and their features is the competence of library science. This could be a periphery sphere of two sciences and all points of interactions in this sphere we can name as integrated factors (Figure 1)

Integrated library factors

Integrated factors of today's libraries are: library functions, library technologies and information media.

1. Library functions

Nowadays libraries perform some main and accessory functions. The main library functions or directions of performance are: user

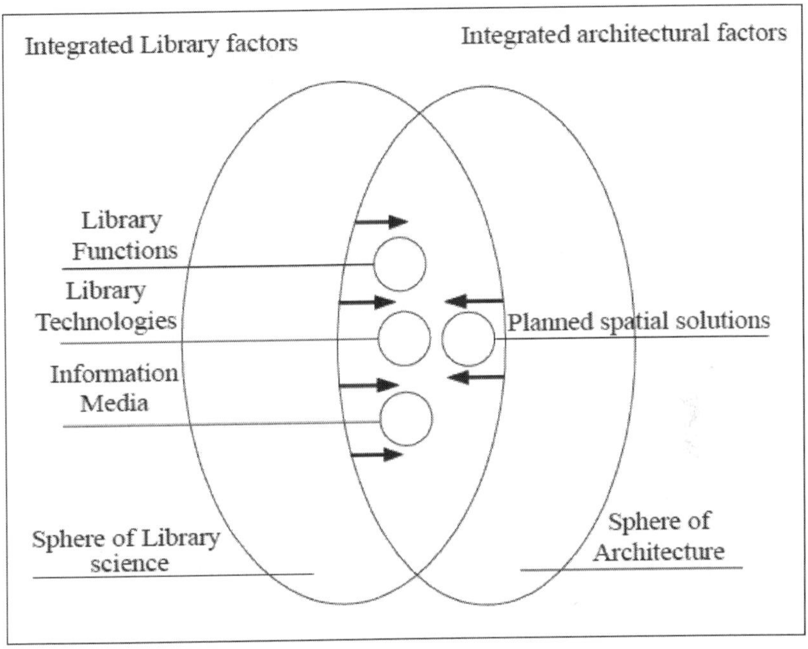

Figure.1. Interaction of integrated factors of library science and architecture

service, media processing, preservation, and organisational-service functions. These main functions are typical for all kinds of libraries. According to these functions library space is divided into three parts: space for user services, space for library media, and space for the library staff. For a long time most of the library services have been oriented on the average user which is not existing in reality. Now libraries are more and more treating users as individuals with very personal needs, approaches and psychological characteristics. Libraries have constantly analysed their services for the purpose to assure that there are enough users who need such services. Only orienting to its users needs, implementing the most innovative forms of media preservation and provision, processes of modern library technology, analysing all these phenomena, determining certain procedures and foreseeing main directions of its performance, libraries can become modern and attractive institutions that can meet informational, cultural, communicational and other needs of the society.

2. Library technologies

Library technologies are one of the most important factors in planning the structure of space for user services, media preservation and staff, and connections and links between these spaces. This factor determines the usage of computer, audiovisual and other modern technologies in the library. Three major technological processes can be distinguished in the library performance. Sometimes these processes are called "library cycles" or "ways". These are: "way of the media", "way of the orders", and "way of the information". The "way of the media" is the most important library circle and includes processing and location of the media. The "way of the orders" includes identification of the media in the information system and in the stacks, delivering media to the users and returning it back to the stacks. The "way of the information" includes the producing of the "secondary" or processed information. All cycles are divided into smaller structural parts - work processes and smaller operations.

In analysing library technologies it is very important to understand and examine flows of technology and materials. These two flows impact differently on the planned-spatial structure of the library. Flows of information media, users and orders are material and can be characterised by length, speed, direction, continuity and discretion. Technological flow is a gradual change of different tools, states and qualities of library work, realised in library circles and processes. The importance of dividing technological and material flows is emphasised analysing the planned-spatial structure of the building. These differences can be solved in two ways: in the first case one can minimise the length of the flows and number of the medium service points with the help of architectural solutions. In this case the flow of media and users is material, but work what is done at the medium service points (carrying and sorting media etc.) is technological. In the other case the way of the user could be shortened using modern library technologies.

The continuing process of development of library technology, based on modern information and communication technology, has allowed planning of library buildings by taking account of future technologies. Otherwise, beside the purpose is to plan such buildings where constructional imperfection can be compensated by sophisticated technology. For example: the storages have to be planned in such way that optimal climate regime can be maintained either by special conditioning system or without it, ventilation can be either artificial or natural. Planning of the building and technological solutions have to be rationally coordinated. Talking about the impact of modern information technology on library buildings planning first of all one has to take account of the usage of computers in library's work.

3. The form of information media

The form of information media is a very important library factor and has direct impact on the structure of storage-spaces. It varies very rapidly now. The technological progress has a crucial impact on this variation. Despite of this, the biggest part of Lithuanian libraries' collection consists of traditional forms of media - books, magazines and newspapers (Kostinaitė, 2003). Modern forms of information media such as microfiches, compact discs etc. are comparatively rare, because of the lack of special equipment for its usage in the mentioned Lithuanian libraries. At the Lithuania National Library only about 0,1% of modern information media form part of all preserved media, including discs, cassettes, tapes, microfilms, microfiches and compact discs. This percentage is even much smaller in the other two analysed libraries. However, according the acquisition reports of these libraries, this percentage is growing every year.

Planning the library storage-spaces, it is important to forecast the changes of the forms of information media and the speed of growth of their amount in the future. The size of the library collection has impact not only on storage-spaces and their qualities, but has impact on all sizes of library spaces as well. Having a rich collection, a library can expect more users and needs more space for reading places. Other integrated factors such as organization and coordination of library performance have not very much impact on architectural solutions or this impact has an indirect character.

Integrated architectural factors

Every typical group of buildings is identified according to the adequate features and characters of architectural solutions. Analysing integrated architectural factors it is important to discuss general characters and future developments of planned-spatial solutions of libraries as a certain kind of buildings.

1. Planned-spatial solutions

Today the planned structure of libraries is not very sophisticated. Very often they have rectangular (close to quadrate) forms. Planned structure of libraries is characterised by three generalised structural elements: area for user service, storage-space for information media, and area for the staff. There can exist one sided or mutual connections between these areas and it is most advisable to analyse them together with all their elements. Every structural element according to library practice worldwide can have many solutions (Library, 1997). Analysing the planned-spatial solutions of the world's and Lithuanian libraries many are massive in size close to the form of rectangular prism and distinguishing themselves by their monumentality. Very often it is one single bulk, sometimes with an internal yard.

Reconstruction or addition of rotund forms of buildings is very complicated, so there are not so many rotund libraries (planned) and in most cases such libraries are comparatively small. For example: Mito Vest City library (Mito, 1993) or V™rmd\ municipal library (V™rmd\, 1990). The rotund form of the Mito Vest City library (floor space 2570 m2) does not perform archival functions. This library has no closed storage-spaces. The whole library collection is in open stacks. The library performs mainly informational, cultural and aesthetical functions, while educational or research needs are insufficiently satisfied. It has a constantly changing collection of new periodicals and publications for leisure and relaxation (Mito, 1993).

In most cases library buildings have two to four floors. Sometimes libraries have just one floor or, on the contrary, many floors (Bibliotheque de France). According to research, this is determined by architectural townscape environment. Most library facades are smooth, clear and strict, with much glass. Some library walls are completely of glass. The walls of the Setaka Public library (Public, 1997) are transparent even from three sides. It depends on the requirement to illuminate spaces for the users. Moreover, it helps to create the image of a modern library - an open and democratic institution, serving the society.

Summarizing, we can say that most bulks of modern library buildings are massive, monumental; in most cases their forms are simple rectangular prisms, with smooth, clear and strict facades, and much glass. It can be seen as a semantic code of the library. Variations are possible only in the frames marked by this code. The concrete natural, urban, and architectural environment influences these frames.

Library Structure And Character Of Its Premises

Premises for user services - reading places, lending and other information media delivery points - occupy 15% to 45% of the total library space. Modern premises for user services can be characterised by one large open space with reading places, lending and other information media delivery points - these points are only different areas of this total space. Internal walls are reduced to a minimum. Very often they are designed from transparent materials. In the library practice worldwide it is difficult to find very strictly separated spaces for information media storage and users. Part of the collection, which is intensively used, is located in the area of user premises. In most of the libraries worldwide the biggest part of their total collection is located in open stacks. The staff service points in open stacks or in lending points are computerized. They have direct connections with the entry area, reference services and catalogues.

Closed or semi-closed working places are rather popular in libraries. According to the above mentioned survey results, 70% of users would like to work in more isolated areas. Wide usage of computer technology does not change much of the planning of user space, with exception on the size of readers' tables, which are planned much larger than in traditional libraries. Some changes in planning user spaces can be mentioned in the catalogue area. Increasing usage of electronic catalogues allowed planners to reduce the catalogue area.

Special premises for audiovisual equipment (for use of video films, compact disks, magnetic tapes, records etc.) are often planned in the users' area. There is increasing need for multiplying, copying, scanning and printing equipment in the libraries. This equipment makes library work more effective but needs special space. This equipment has to be easily reached by the users. All analysed Lithuanian libraries lack such equipment. Most of their equipment is out-of-date and such services are expensive.

Some decades ago information media storage-spaces have been designed for preservation of books and other media published in paper form. Nowadays, because of the rapid growth of printed materials acquired by libraries and because of the varying forms of information media the planning of storage-spaces and its equipment has experienced many changes. During the last years more and more expensive, fully automated storage systems are being used in many libraries of developed countries. There also exist not very expensive and simpler storage systems in other libraries as well.

Reviewing professional publications, we can meet the stress on future storage problems. We can predict the continuing growth of various forms of information media, implementation of high information technology in the library that will change the structure and shape of storage-spaces. We expect that the space for storages will be reduced compared with other library spaces. In current libraries spaces for storage occupy 15% to 50% of the total building. Planning future storage-spaces one must predict the possibility to expand these in horizontal and vertical directions depending on the total composition of the building. Space for staff premises is not specific for library buildings.

Nowadays these are usually computerised and automated. This space occupies 7% to 30% of the total library space and in the future this percentage will be reduced, because of the further computerising and automation of library processes. Auxiliary premises (lobbies, lavatories, locker-rooms etc.) are not specific for library buildings. They usually occupy 10% to 20% of the total library space.

Some Conceptual Principles Of Planning Modern Libraries

The traditional library doesn't meet the changing needs on information provision of today's information society. Library buildings, which are planned on old norms and standards, do not satisfy today's situation. Summarizing the results of above-mentioned research, we can formulate some principles of modern library's activity and its architectural solutions.

1. The modern public library has to be multifunctional

The modern public library has to be multifunctional and to provide not only informational or other typical library services but cultural, communicational and other extra services as well. Such library will comprise two functional parts: 1) provision of library's traditional (informational) services and 2) cultural and social services. These non-traditional services (cultural, educational, commercial, sports, recreational, entertainment etc.) will make the library more attractive and can lead to an increase of the number of users of the traditional library services. For this new cultural and social functional part adequate premises have to be planned. This will give the library a new image of social importance. The modern public library will become a social centre of culture, a home for society. It will be a multifunctional and vital centre, organising and joining society. A library with such social and cultural activity will be more attractive for investments. The library will become a library information centre, club, and Internet café etc. Oriented to the various needs of society such a library will satisfy needs of different social groups and ages. It will become a synthesis of various cultural activities. All these different kinds of activity will stimulate other kinds of cultural activities and all kinds of culture in general. So, the public library has to be perceived as an information institution and as an amateur art centre and organizer of recreation and communication as well. According to the results of the mentioned survey and research on integrated factors, we can state that 20% to 35% of library premises have to be assigned to social and cultural activity, 50% to 65% for library's information services, 20% to 25% for storage-spaces, 5% to 10% for staff premises, and 10% to 20% of space have to be assigned to auxiliary premises (Fig.2).

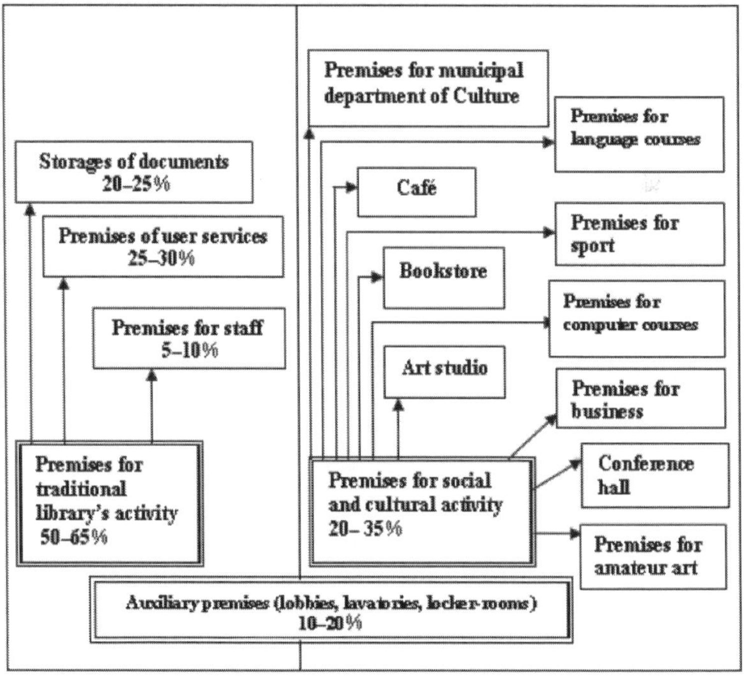

Figure 2. Model of multifunctional public library

2. The modern public library has to be comfortable

Comfortableness of service is one of the most important indicators of modern library activity's value. This indicator includes: easy access to the library building, rational planning of the building and impressive architecture. There should be created a good working environment in the library: arranged individual carrels for intensive working and studying users, and well illuminated working places with dividing screens separating each place. Comfortableness also includes: providing the library with soft furnishing, plants, aquariums, soft carpet, large windows, balconies, premises for communication of people with different interests, and provisions with sound and video equipment. There is one more important demand for comfortableness. This is the accessibility of the collection. The collection has to be located according to library's collection profile, library's strategy and expectable users. The whole collection, arranged in subject groups, has to be located in different areas of open stacks.

3. The modern library has to exist in harmony with its natural and urban environment

The library has to become an organic part of the natural and urban environment.

4. The modern library as an open, democratic and intellectual communication institution

All architectural solutions of the modern library have to express the image of the library as an open, democratic and intellectual communication institution. This image can be created by planning buildings of "virtual" architecture what can be characterized by "dematerialised" forms of clear, laconic, simple and visible bulks. In the process of making constructional resolutions and choosing the building materials one has to orient on comfortableness of the internal and external environment of the library. There has to be a balanced thermal, illumination and noise regime. The building of a public library has to be not higher than four overground floors. The higher a building, the more expensive will its maintenance be, more staff will be needed and more sophisticated technology and security systems.

5. Planning of the building has to be flexible and simple

The planning of the building has to be flexible and simple in the general plan. There have to be possibilities of extension or transformation. That's why is advisable to use as much as possible movable columns, equally located ventilation, heating and aeration systems. The construction of light screens should also be easily movable from one to another to create a large space.

6. An internal garden in the middle of a building

It is advisable to design an internal garden in the middle of a building. This solution together with vertical and horizontal cabling allows an architect to create an optimal and aesthetic interior. Through the internal garden it is possible to achieve an architectural wholeness of the building and the interior.

7. Locate the storages on the underground floors

It is advisable to locate the storage-spaces on the underground floors and premises that need daylight on the overground floors. The underground storage-spaces can be characterized by the following advantages: the possibility of rational usage of the collection, to locate it operatively, to organize more credible security of the collection, and make it much easier to maintain a proper temperature and humidity regime. The underground storage-spaces make it possible to design a compact building, and better its constructional, economic and exploitative indicators.

Application Of The Main Conceptual Principles In A Pilot Project

For this pilot project (planning of a modern regional public library) has been chosen a plot in the Pilaite residential area of Vilnius city. This area has a good transport connection with all areas of the city. The library is planned in the very centre, on the compositional axis of this area in front of a church. All libraries' walls will be of glass. The library will look as a glassy heart of this area - open to all world's news, information, and problems, and inviting to visit. The shape of the building is very simple - a bulk of the quadratic projection. One of the building walls is concaved inside. It looks as if standing opposite to the church influences it. This arched line adds variety to the exterior of the building and gives it an expressive character.

The building is planned in the centre of the plot, and has very convenient pedestrian tracks and roads to parking places. Near the library is planned a square covered with stones, which contrasts with the building's glassy facade. In this square are planned relaxation places and one can see reflected on the glassy library walls a composition from stones and trees. The bulk of the library is monumental but it looks light and open. It is a structure of virtual architecture with actually dematerialised outside walls. It looks as if there is no building: it melts with the background of sky and the surrounding of trees and stones (Fig. 3).

Figure 3. Model of the Building

On the roof of the building there is a glazy shade over a winter garden as if it crowns the building with its round form. It echoes with the concave wall of the main facade. For the construction of the library use will be made of a columns system (12 x 12 m). This allows the architect to act more flexible during the planning of the library premises. This system of columns will be used for the underground floors (Fig. 4.)

The interior of the building is planned rationally: there are many possibilities to transform the spaces dividing them with the help of partition walls. Only constructions of staircases and lifts are inserted into the main carcass of the building. There is a winter garden inside the building covered by the glazy shade. Through the garden's glazy roof daylight comes into the building and through various galleries it diffuses to reading places, lobbies and other premises. The garden is the heart of the building. It's like the building's core with a game of water and lights on the "water wall", fountain, playful aquariums and decorative plants.

The total area of each floor is 1200 m2, with total library's area of 6000 m2, 4 overground floors and 1 underground floor. On the first floors are located a lobby, locker-rooms, a children's reading room, a café-newspaper's reading room, a bookstore, the library acquisition department, the winter garden, and premises for cultural activity (Fig.5).

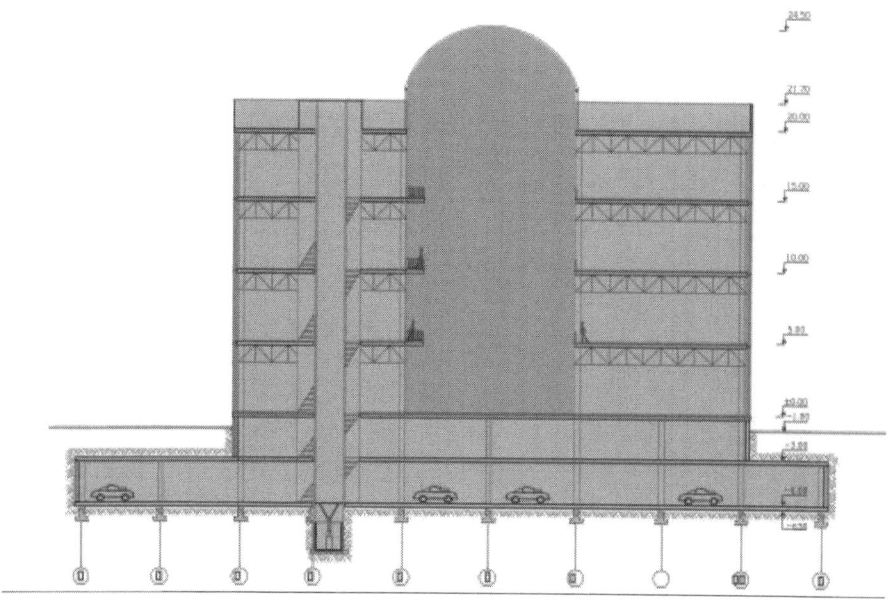

Figure 4. Cross - section of the building

On the second floor: open stacks with the fiction's collection, a silent reading room of current periodicals, an Internet and information searching room, the acquisition's processing room, the director's office and a meeting room, a gallery, an arts studio, and space for group work.

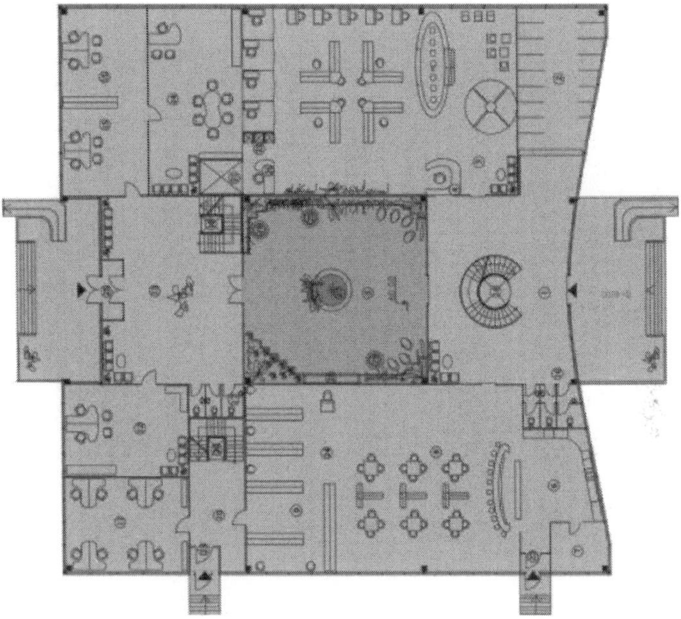

Figure 5. Plan of the first floor

On the third floor: open stacks with the Humanities collection, a silent reference reading room, an information searching and usage room, an information preparing room, a gallery, premises of discussions' club, and space for group work (Fig. 6).

On the fourth floor: open stacks with the Technical literature collection, a conference room, the reference department, a gallery, space for amateur art activity, and space for group work. On the underground floor is planned closed fully automated storage, and space for sport activity.

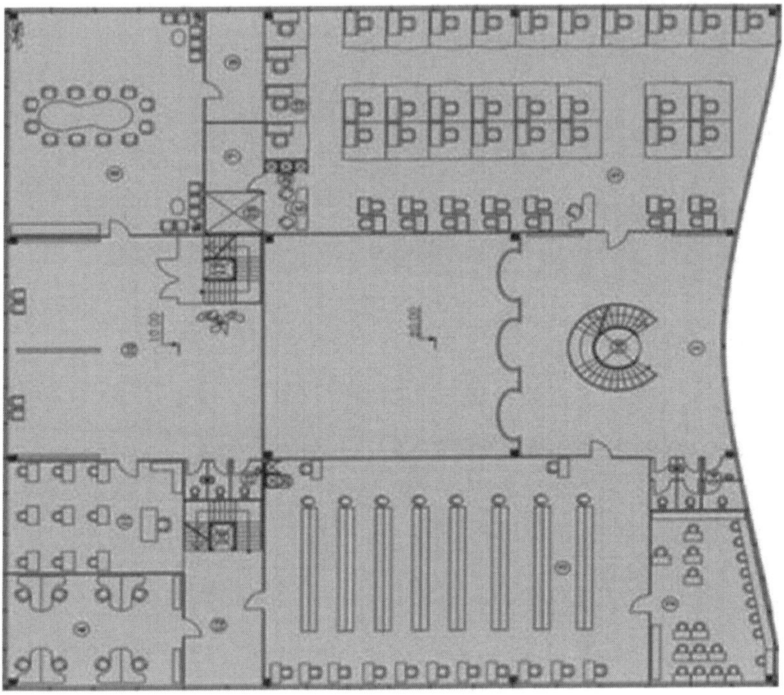

Figure 6. Plan of the third floor

The planned building justifies the results of the above-mentioned research and illustrates the validity of the formulated conception of a modern library. The proposed architectural solution has no pretensions into the final decision; this is only one possible way. In each concrete situation its urban and natural environment influences the architectural solution of a library.

REFERENCES

1. Jakovlevas-Mateckis K. *Kai kurie miesto aplinkos sisteminio formavimo aspektai*. Statyba, 2000. Pp.:27-33.

2. Kostinaitė L. *XIX-XX a. pasaulio ir Lietuvos viešųjų bibliotekų architektūros apžvalga.Technikos mokslų raida Lietuvoje.* Mokslo istorikų konferencijos pranešimai. Vilnius, 2001. Pp. 46-54.

3. Kostinaitė L. "Šiuolaikinės viešosios bibliotekos kai kurie architektūros ir veiklos bruožai". In: *Šiuolaikinės architektūros problemos.*. Vilnius, 2003. Pp. 95-106.

4. Kostinaitė L. and K. Jakovlevas-Mateckis. "Šiuolaikinės viešosios bibliotekos architektūra: raidos tendencijos, jai turintys įtakos veiksniai, kai kurios projektavimo nuostatos". *Urbanistika ir architektūra*, 26(2002)1, 37-48.

5. *Library Builders.* London : Academic Editions, 1997. 224 p.

6. Mito Vest City Library (GIRO). *The Japan Architect*, (1993)1, 144-146.

7. *Šiuolaikinės Architektūros Problemos.* Konferencijos pranešimų medžiaga. Kaunas : Technologija, 2003. 107 p.

8. *V™rmd\. The Library of the '80s.* Stockholm,1990. Pp. 55-59.

9. Public Library Setaca Project. *The Japan Architect*, (1997)2, 174-175.

4

CHAPTER

Cabriology: Putting Digital Collections in the Local Flow at Bruges Public Library

Koen Calis

Librarian, Bruges Public Library, Kuipersstraat 3, 8000 Brugge, Belgium

ABSTRACT

Cabrio, the Aquabrowser implementation at the Bruges Public Library, is more than a catalogue makeover. Rich, variable, flashy and truly inspirational, its open structure tackles several strategic issues related to a changing network environment. Digital collections are merged with the catalogue to become part of a hybrid library mediation model. Cabrio focuses on local dynamics and existing participation, which are key features of public libraries, in order to challenge and redefine library perceptions and services.

Keywords: Cabrio, public library, digital collections, Aquabrowser

INTRODUCTION

The Bruges Public Library is a lively public institution engaging with the local community as well as a special library harbouring rich heritage collections. This is reflected in a wide range of users. We have tried to take advantage of this very complexity. Digitisation and access to digital collections should be part of an overall model: a rich hybrid local hub connecting a broad audience to inspiration and knowledge, encompassing old and new, inside and outside, local and global, specialist and general needs.

As a public library we are well aware that the environment we work in has changed and that digital libraries are as much about users and connections as they are about collections. The traditional library catalogue has increasing difficulty to fulfil the expectations of the average user of, say, Google, Last.fm or Facebook, who is used to get immediate and often unintentional access to a wealth of content and connections he can control. As a result of this, new forms of participation and (collective) authority arise, which run counter to the traditional communications model of the public library. These are issues libraries should not solve on their own. But it is interesting to analyse how an individual institution can reconcile a collective development with exactly that local focus and commitment that makes a public library so unique in its physical presence.

That is where Cabrio comes in. This Aquabrowser implementation at the Bruges Public Library is more than a catalogue makeover. The acronym, in Dutch reminiscent of a flashy convertible car, connects 'catalogue' with 'Bruges', 'inspiring', 'internet', 'discovery' and especially 'openness'. In its richness, its variety and truly inspirational nature, its open structure tackles several strategic issues related to a changing network environment. It focuses on local dynamics and existing participation, which are key features of public libraries, in order to challenge and redefine library perceptions and services.

INTEGRATION

Overcoming Internal Fragmentation

The starting point was the fragmentation of local discovery. In addition to public library materials the library owns rich heritage collections ranging from medieval manuscripts and old printed books to personal collections, among which are the important literary archives of the poet Guido Gezelle. Information on these collections was dispersed over three databases and data formats:

- the public library collections and special collections from 1830 on (MARC concise)
- the Guido Gezelle archives archival database (proprietary format)
- the full MARC database for the old printed and manuscript collections.

While public library users had to deal with an alphabetical sorting of results containing a lot of items from the older collections, the two heritage databases focused mainly on expert users.

Furthermore the library had started digitisation, e.g. of a frequently used genealogical manuscript, a rare vernacular manuscript of the Liber Trotula, a medieval treatise on women's medicine, and Colard Mansion's incunable Ovide Moralisé. These projects are part of a

cooperative local digital heritage platform http://www.historischebronnenbrugge.be. Each resource is well documented and includes specific retrieval features (e.g., a name index for the funeral manuscript or an index on mythological themes or woodcuts for the Mansion incunable), which give them an added value and make them searchable through Google.

Reconciling preservation with participation was an important objective for the digitisation of the brittle Gezelle archives. The complete poetry and photo archive can be consulted digitally. In cooperation with Antwerp University, there an inventory of the letters was made and several correspondence editions are available online. An educational website, http://www.gezelle.be, attracts users from Belgium and the Netherlands and leads them into the database. Finally, full in-house digitisation of all medieval manucripts has just begun.

All of these collections are now available through one integrated interface in Cabrio, opening them up to a variety of users. The digitised funeral poetry of Guido Gezelle for instance is an unexpected source of nineteenth-century neogothic printing in Bruges. In cooperation with the Bruges City Archives a pictorial database was added, which blends well with the original content. Integration of eighteenth- and nineteenth-century newspaper and almanach collections is on the way. A search on the medieval abbot Idesbald reveals eighteenth-century old printed devotional books as well as poems, pictures and documents relating to his late nineteenth-century sanctification, actually telling his story through the ages. As a result of this, focus has shifted from access to a topical interest. Which resources are missing? How can the library connect these heritage threads to the interests of people or to scientific research?

EXTERNAL SOURCES

From a public library perspective Aquabrowser allows the library to plug into economies of scale and collective library development without losing its strong local focus. The Bruges Public Library has a long-standing tradition in cooperative cataloguing, but a lack of integration often prevented the Library from reaping the full benefits. One of the key issues in Cabrio was vertical integration with the facilities organised by the Flemish Government through Bibnet, the Digital

Library Agency, especially the national public library hub bibliotheek.be. The databases are now being matched and merged on a daily basis. Among the benefits are data enrichment (merging of specific fields, covers, articles and tracks), thesaurus integration, the sharing of digital collections (websites, newspaper articles, etc.) as well as cooperative rights and cost negotiations. This catalogue mash-up also makes it easy to adopt innovative services developed centrally, such as the digital lending of music, the implementation of relevance ranking or the last fm tips. The platform is in place and is at this moment being implemented as a Flemish model for library networking (Bibnet Library Portals project).

Secondly, the library catalogue is complemented by external information resources, mostly local or literary. At this moment about 15% of the content is immediately available in digital form, and accessible for home use. The aim is not to compete with search engines, but to offer a selection of valuable digital content along with the physical library collection with the local audience in mind. The contents of a literary blog 'De papieren man' and the important Digital Library of Dutch Literature (DBNL) are being crawled. If you search for the poet Christine D'haen you get the news of her recent death or biographical details and essays together with the library content. The record detail also has external links to LibraryThing, subscription databases etc. The addition of In&Uit Brugge, local events and courses, as well as community information crawled from the web integrates the live environment within the library walls. In this way, a child looking for 'planets' is linked to the Bruges observatory. The query 'dementia' not only results in books and articles, but also links to local organisations, lectures and discussion groups. A search on 'Joss Stone' reveals her presence at the 2009 Bruges Cactus festival.

Local 'anchoring' of the library is crucial, because the combination of vertical (sectoral, general) and horizontal (local) integration makes the library catalogue a unique focal point, combined with an accessible cultural and knowledge agent. Interestingly, potential partners who find themselves mirrored now within the library walls, have taken an interest in the library as a channel for reaching a broad audience of 40,000 direct users. It is a win-win situation, because the library is able to transcend the traditional media and delivery services that are so closely connected to its identity. For many people, the library is an a priori choice for or against a medium or the lending function. Perception rather than knowledge orients this decision. By enriching the library and making it hybrid we hope to shift the focus from content to guidance and user needs. The choice of a specific medium should be dealt with, subsequently, through refinements in the system. This concept of the library as a place of discovery was effectively translated into a promotion campaign. It showed people who ended up in unexpected life-enriching situations as a result of their catalogue search.

Creating Value: What's in it for Me?

The library's second main ambition was to develop a mediation model for this hybrid content, guiding users not to a blob of information, but only to what is valuable to them. The fine maze of physical local presence, the trusting relationship with users, the expertise of the staff, the knowledge of the local field and the library tradition of structure and quality are assets that distinguish the library from other actors in the new context.

SEGMENTATION

First of all, we felt it was important to reduce the overload that is the result of the richer library. That is why Cabrio is segmented in subsets, indicated by tabs. As the ubiquitous records of the stack collection hindered the use of the open shelf collection, we decided to put the closed stack collection together with the heritage collections in the last 'heritage' tab. The open shelf collections are divided into an overall search and four subsets: children's library, fiction, non-fiction and music&film. These tabs also allow us to build focused search and discovery environments. Each domain has its own internal and external data, its own refinements and its own showcase with cover animation, menu and blog posts. This means that someone who is looking for novels in 'fiction' will not only find an inspirational reader's advisory environment, but he will also get more focused results not hampered by irrelevant articles, tracks or other content that happened to come within the scope of the search term.

Smart Searching and Sorting

The search is enriched with pseudonyms and synonyms. Related thesaurus terms and associations based on the data are offered in a word cloud. Of particular importance is the Flemish Bibnet thesaurus which we integrate locally and which operates across the various resources. This means that a query such as 'cookery' also will result in web crawled pages in which the term 'cuisine' is mentioned. In order to increase the catch, stemming is applied to some fields which takes account of word derivations and compositions. Much attention was paid to ranking the results, with the most relevant or most recent items listed first. Relevance ranking takes into account the number of items in possession and the distribution, so that popular works are higher in the ranking. Furthermore, specific field indexes such as subject or media citations are sorted by a combination of year of publication and acquisition date. The top results are always very fresh and keep changing as the collection develops.

Discovery through Rich Browsing

On the other hand, we want to give users a reason to use the system which goes beyond delivery, by connecting them to content they had no prior knowledge of or interest in, or for which they could not formulate the need.

To attract an audience beyond the real 'information seekers' we felt it was important to appeal to users not only in a literate way. The inclusion of covers in the results list and the detail page is not only aesthetical, but has an advisory role in the selection process. Cabrio also includes public domain trailers for novels and movies or podcasts of authors. We can easily conceive this for new fiction or information books, but it also applies to heritage collections. Imagine 25 editions of the same poetry book by Gezelle. Looking at the colour and the typography of the cover is an important means to distinguish between them.

As statistics show, one third of the queries are not search box initiated. This means that people follow the leads and choices that the system presents them. People use the system in a similar way to search engines. 90% of the searches stay on the first result page, either clicking through to the next first page or refining them. It allows users to develop previously vague needs on the go, very similar to the way they use the physical library. In order to do so, the system has been designed specifically to exploit the rich metadata structure, giving users rich browsing possibilities, including an attractive clickable map, or giving them smart refining options. Specialised advisory data such as media references, ranging from women's magazines to canonical literary guides are added. These reflect the diversity in life styles and media preferences. The same field is used to connect objects to local activities. With just a few mouse clicks it is possible to find Scandinavian thrillers in translation that were recently discussed in the media or historical novels shortlisted for the Booker Prize owned by the library. Heritage collections can be navigated through or the searches refined by provenance, publisher, exhibition, acquisitions etc. The exhibition field is an interesting one because it not only reflects the educational activities of the library, but also highlights the most precious collections or the major themes. For instance, one can easily see which medieval manuscripts were featured in the prestigious 2002 'Closed World Open Books' exhibition. 'The epitaph on paper' exhibition provides a thematic pathway to manuscripts containing funeral inscriptions collected mostly in the French period. Contrary to large-scale digital libraries that concentrate on a general common core of metadata, we feel it is important for us as a local institution to provide specialized data that describe the different contexts heritage objects are part of and open up a fabric of choices for rich browsing.

Showcase with Integrated Blog

The data can be accessed both from the individual record and from canned searches in a front end menu. This menu is part of the showcase of each segment, together with a cover flow animation of acquisitions and an integrated blog. The blog entries aim to activate content, connecting it to what happens in the library, the city or the broader environment. It is a very useful tool to bring long-tail content to the user's attention. Alternatively library activities such as an important acquisition or the annual children's literature festival finally get a face. It is very satisfying to see the lively physical library mirrored in the digital one. Because the blog is a prominent feature of the virtual access to the library, it is not a random exercise. On screen and on an annual basis, we seek representativity. We aim to post at least one entry for every tab per week, which means a full refresh of the screen within a loan period of four weeks. To plan this, we lay the library's internal calendar next to the agendas of the various socio-cultural partners. In return, the activities in Cabrio or new partner synergies give a stimulus to the physical library, creating new ambitions and work processes.

Connecting with the Community

The blog functions independently, but has Cabrio as its primary channel. It is a good example of how innovative services are embedded within the traditional environment and user activities. Users who wish to respond, are referred to the original blog. About 15% of blog entries receive one or more responses. Whether that actually happens is very much dependent on the type of message. Closed announcements often get no comment. A blog entry about a poster contest peaked at over 100 responses. Editorially we try to respond to this by writing posts that are more inviting to conversation, such as the list of ten novels every tenth day of the month to which users can add their own. Each response is again linked to the catalogue. Experience shows that the blog is beginning to lead its own life. Users subscribe to it. Posts are quoted or commented on, leaving traces outside the library and making them a privileged object of search engines. Within the organisation itself blogging soon leads to the use of related services like flickr, slideshare, youtube etc. which in turn activate content extra muros. Library use naturally overflows into outside networks. The same happens with the deep LibraryThing link or the Add-This button. In December 2009 Aquabrowser MyDiscoveries was added, allowing users to select, share and interact across library boundaries. Mobilising content has become a major issue.

Cabrio has caused a significant rise in home use of the library's own channel. With an average of about 8,000 unique external visitors per month we estimate that about 40% of the physical library users use Cabrio at home. They also use it more frequently than the previous web-based catalogue. Other successful examples are the above-mentioned educational websites www.historischebronnen

brugge.be and http://www.gezelle.be. Both are frequently being linked to by Wikipedia and other sites and they attract an audience that goes beyond expert use. Moreover, a search tree has allowed search engines to index the Gezelle archives inventory at item level. And yet, there is a lot of room for improvement when it comes to proactively reaching out to non-library user environments to enable users to discover and reuse library content. Widgets exist and RSS is being studied. Obviously, the initiative lies with the user and the library has to show itself rich enough to be worthy of the attention. In this respect, the hybrid content as well as the rich structure and advisory data are an important basis on which to build future services. As far as local institutional partners are concerned, rather than focusing on a commercial business model, we are trying to develop reciprocal arrangements, both in content and in the channeling of attention. At the moment outside-in linking happens mostly on a macro-level to Cabrio as a whole, but we are trying to develop more in-depth connections. What matters is to find a place on the radar of the users at the time when they have a certain need. Again the vision is to focus on cooperation with partner institutions at a local level, alongside with negotiating flows collectively as a library network.

ACKNOWLEDGEMENTS

The main structure of this paper is based on Robin Murray, 'Library Systems: Synthesise, Specialise, Mobilise', Ariadne 48 (2006); http://www.ariadne.ac.uk/issue48/murray/.

5

CHAPTER

Public Libraries – Facilitators of Information Society and e-Inclusion in Latvia

Signe Balina

University of Latvia, Faculty of Economics and Management,Aspzazijas blvd. 5, Riga, LV-1050, Latvia

ABSTRACT

The economic growth and national competitiveness is closely correlated with the development of information society. In the last few years the role of public libraries has significantly increased and libraries have greatly contributed to the improvement of information society in Latvia. The usage of library services had never been as high as they were in 2011 – 42% of the population were active library users and 90% of them used the Internet. Nowadays, the public library network in Latvia provides a powerful network of information, knowledge and community centres and they are highly demanded by society.

Keywords: information society, economic growth, e-Inclusion, public libraries;

1. INTRODUCTION

Nowadays, the economic growth and national competitiveness is closely correlated with the development of information society. E-Inclusion is one of the most important cornerstones of the development of information society in order to be sure that all social groups are included in the reducing of the digital divide. In Latvia, e-Inclusion is fostered by a number of stakeholders representing different sectors. The successful development of information society in Latvia lies in the fact that all of more than 800 public libraries are connected to the Internet; all librarians are trained to use information and communication technologies (ICT), as well as trained to help people work with computers and search for relevant information on the Internet or in databases. The great advantage of libraries is their ability and willingness to collaborate with all stakeholders in order to reach all social groups.

2. THE DEVELOPMENT OF INFORMATION SOCIETY AND ECONOMIC GROWTH

The usage of ICT is one of the main drivers of change within society and business for more than a decade. An inclusive European information society is one of the priorities of the strategic EU framework i2010 – a European Information Society for growth and employment as well as one of the pillars in Digital Agenda – the flagship initiative under the Europe 2020 strategy to deliver smart, sustainable and inclusive growth.

There are several indicators that characterize the development of information society, such as percentage of individuals using the Internet, number of households with Internet access and number of households with broadband access. The economic development, however, is characterized by both gross domestic product (GDP) and national competitiveness.

The Global Competitiveness Report 2012 – 2013 allocates countries (economies) into three stages of development: 1st stage - factor-driven economies; 2nd stage - efficiency-driven economies; 3rd stage - innovationdriven economies. Any countries falling in between two of the three stages are considered to be in transition. (Schwab, 2012)

There is a strong statistically significant linear correlation between the percentage of Internet users and the natural logarithm of GDP which confirms, as previously mentioned, that the usage of ICT is one of the main drivers of change within society and economic development.

From the graph we can see that the percentage of Internet users differs for countries in different development stages. This finding is also supported by the results of the analysis of variance which shows that the average percentage of Internet users differs between all five groups. There are four statistically significant homogeneous subsets divided by the average percentage of Internet users which indicate also four homogenous groups of countries with different levels of development of information society (see Table 1).

Table 1. Average percentage of the Internet users and development stage of economy

Development stage	Number of countries	Subset for significance level $\alpha = 0.05$			
		1	2	3	4
Factor-driven (Stage 1)	38	9.9			
Efficiency-driven (Stage 2)	33		33.0		
Transition from Stage 1 to Stage 2	17		33.3		
Transition from Stage 2 to Stage 3	21			57.5	
Innovation-driven (Stage 3)	35				76.7

Factor-driven economies have the lowest average percentage of Internet users - 9.9%, as well as the lowest GDP per capita. Innovation-driven economies have the highest average percentage of internet users - 76.71%, as well as the highest GDP per capita. There is no statistically significant difference between Internet usage for efficiencydriven economies and for countries in transition from factor-driven to efficiency-driven.

Factor-driven economies have the lowest average percentage of Internet users - 9.9%, as well as the lowest GDP per capita. Innovation-driven economies have the highest average percentage of internet users - 76.71%, as well as the highest GDP per capita. There is no statistically significant difference between Internet usage for efficiencydriven economies and for countries in transition from factor-driven to efficiency-driven.

3. E-INCLUSION – AN IMPORTANT COMPONENT OF INFORMATION SOCIETY

The Latvian economy is in transition from efficiency-driven to innovation-driven. There are 71.7% of individuals who use the Internet in Latvia. The Internet usage level is much higher than the average of the corresponding development group: for efficiency-driven economies this percentage is 57.5%. The analysis of the Latvian GDP growth rate and the percentage of Internet users shows that the proportion of Internet users increased even during the period of recession in 2008 and 2009 – although the GDP rate rapidly decreased, the percentage of Internet users continued to increase, just slower (see Figure 2). Therefore we can conclude that in Latvia aspects other than the GDP had greater influence on the increase of Internet usage

In order to fully participate in society, people now need to be competent and willing to use ICT tools to access information and engage in social processes.

One of the aspects of successful development of an inclusive information society is access to the Internet. If we consider the indicators of Internet access availability and quality in Latvia we can see that the quality of the Internet in Latvia is even higher than in average in Europe – the share of fixed broadband lines equal or above 100 Mbps is 15.5% (average in EU only 1.3%), 3G coverage is 99% (in EU – 89.9%). At the same time we are below the Europe's average number of households with Internet access and number of people who can access the Internet through a mobile phone (see Table 2). There are several reasons for this. First of all, we have insufficient Internet access coverage in rural areas; secondly, not all households in Latvia have a computer. The latter is economic factor, but here a significant role plays access to Internet secured by public libraries in Latvia.

Table 2. Access to computers and Internet in Latvia and EU, 2011.

Indicator	Latvia	EU 27
Share of fixed broadband lines equal to or above 100 Mbps	15.5	1.3
3G coverage (as a % of total population)	99.0	89.9
% of population accessing the Internet through a mobile phone via UMTS (3G)	3.7	11.8
Households with Internet access at home, %	63.6	73.2

Source: European Commission (2013)

Another aspect is digital literacy. In the last few years in Latvia we have had significant progress in digital literacy. The percentage of individuals with medium or high Internet skills increased from 18.7% in 2005 to 60.9% in 2011 (in comparison to the average in the EU - 22.3% and 42.6%). Latvian Internet users are one of the most active social network users in the EU – 55.1% of the population (average in the EU is only 37.8%) or 79% of all

Internet users in 2011, as well as one of the most active in online job seeking – 27.1% in 2011. (European Commission, 2013)

In the context of the development of information society, it is necessary that all groups of population benefit from the use of ICT and that no one is left behind. E-Inclusion means both inclusive ICT and the use of ICT to achieve wider inclusion objectives. In Latvia as in most EU countries, e-Inclusion is fostered by a number of stakeholders representing different sectors: national government, local governments, businesses and non-governmental organizations. All of them collaborate in order to achieve the main objectives of Latvian e-Inclusion policy. (Balina, 2012)

4. THE ROLE OF PUBLIC LIBRARIES IN E-INCLUSION

In recent years people more and more gain from access to the Internet. At the same time Internet access and digital skills inequality expand the digital divide. The challenge of information society is to motivate and educate people from all society groups in the use of new technologies. Libraries provide an important alternative means of access for disadvantaged people, although the neediest people are not necessarily the ones who use library computers. (Kinney, 2010)

Public libraries in Latvia have greatly contributed to the improvement of information society – in the last 6 years public libraries have faced transformation from traditional library operations to the use of modern technologies. The reason why people visit libraries has also changed.

This transformation started in 2006 when the Global Libraries initiative of the Bill & Melinda Gates Foundation was initiated in collaboration with Latvian government and local authorities. The goal of this public library development project «Father's 3rd Son» was to facilitate and promote the use of resources offered by free access to ICT and the Internet in libraries.

Another important aspect is that all librarians are trained to use ICT as well as trained to help people work with computers, search for relevant information and use public or private sector e-services. This initiative has certainly made a strong contribution to the increased impact that public libraries have on the quality of people's lives. This initiative also facilitates Internet and computer usage as well as increase the level of digital literacy. The influence was stronger in rural areas and for people in social risk groups.

Currently all 817 public libraries in Latvia (149 in cities and towns, 666 in rural area) have Internet access and public libraries in Latvia are in high demand as centres of information, knowledge and community.

Usage of library services has never been as high as they were in 2011:

- 42% of the population are active library users;
- the number of library non-users has reduced from 62% in 2007 to 9% in 2011;
- 90% of the library visitors use the Internet. (Balina, 2012)

The increase in these rates is caused by the social and economic benefits that people gain by using ICT, not the infrastructure or librarians alone. Research done in late 2011 confirmed that library users recognized the access to ICT in libraries as a significant service that improves the quality of their lives. 54.2%

of the users admit that Internet in libraries helped them to access the labour market (job search or employee search). For 53.8% it helped to improve their communication with relatives and friends. 45.4% of the users admit that they have saved money by accessing ICT at a library. 45.3% pointed out that Internet at public libraries improved access to state and municipal eservices. (Centre for Culture Information Systems, 2013)

By analysing the habits of library Internet users in the past 12 months the research shows that 84% also take books, newspapers or magazines to read at home, 74% use printing, scanning or copying services, 43% consult librarians in the use of computers, Internet and databases. (Centre for Culture Information Systems, 2013)

The most common answer by 80% of respondents why they use the Internet in libraries was that this is done because Internet in the library is a free of charge service. More than 2/5 of the respondents used the Internet in the library because the library has a pleasant environment (52%) and in a convenient location (46%). 45% of Internet users in library mention also use of other necessary equipment (printers, scanners, etc.). Importance of librarians' assistance and advice was important for 42% of Internet users, 19% said they use the Internet in the library because it is not available elsewhere. Analysing tendencies of the previous years the conclusion is that the role of libraries increase – people more and more use services in libraries and value them (see Table 3).

Table 3. Why people use the Internet in the library (percentage from Internet users in libraries)

Reason	2011	2009	207
Free of charge service	80	73	63
The library is a pleasant environment	52	23	28
Availability of other equipment (printers, scanners, etc.)	45	21	23
Librarians assistance and advice	42	15	21
Not available elsewhere	19	21	18

Source: Centre for Culture Information Systems (2013)

Percentage of inhabitants for whom library is the only access point to the Internet has increased from 18% in year 2007 to 21% in year 2009 and then decrease till 19% in 2011. It means that for those people library is the only place where they can benefit from information society

Libraries also have an important role in the development of e-skills. Library Internet users increasingly use librarian assistance with computer use. Advice relating to the use of the computer in 2011 has been asked more frequently than in previous years – 33% of respondents in 2007, 27% - in 2009, and then in 2011 this number increased to 48% of the respondents. There are two main reasons for this – the trust to librarians and their knowledge and their ability to answer the questions and to provide help to people: 98% of customers were satisfied with service and 94% with knowledge level of librarians. The satisfaction rate in paste years is growing. (Centre for Culture Information Systems, 2013)

This clearly shows how important libraries are for providing equal access to information especially in case when economic situation is changing.

5. CONCLUSIONS

The economic growth and national competitiveness is closely correlated with the development of information society. Latvia as a country in transition from efficiency-driven to innovation-driven economies has showed the importance of the information society and its effect to the economic growth. Latvia has been strongly affected by the economic recession in 2008-2009, but one of the stimulators of the recovery of economics in 2011-2012 has been also the quality of the Latvian ITC infrastructure and society's capabilities to integrate in the information society. Significant contribution to this success has been played by public libraries having greatly contributed to the improvement of information society in Latvia and growth of the economics of Latvia.

REFERENCES

1. Balina, S. (2012). Inclusive Information Society in Latvia. In Proceedings of the 5-th International Scientific Conference on Applied Information and Communication technologies, Jelgava, Latvia, April 26-27, 2012 (pp. 153–156). Jelgava, Latvia: Latvia University of Agriculture, Faculty of Information Technologies

2. Central Stasticial Bureau of Latvia. (2013). Statistics database.

3. Centre for Culture Information Systems. (2013) Latvian public libraries and the Internet: Technology, Services and impacts, 2011.

4. European Comission. (2013). Scoreboard. Retrieved from

5. Kinney, B. (2010) The Internet, Public Libraries, and the Digital Divide. Public Library Quarterly, Vol. 29 Issue 2, pp.104 - 161.

6. Schwab, K. (Ed.). (2012) The Global Competitiveness Report 2012–2013. Geneva: World Economic Forum.

6
CHAPTER

Perceptions on the Effectiveness of Communication between Public Institutions and Journalists through Social Media

Mihaela Păun,

PhD student University of Bucharest, Romania

ABSTRACT

This article presents the challenges and opportunities of social media for public institutions and argues that the designing and implementing government public relations using social media involves more than merely having another communication channel for publics. I will explain the "social media", the differences between social media and electronic-Public Relations - E-PR, and the communication between public institutions and journalists. The interviews with journalists or with PR specialists in public institutions were focused on the use, the extent of this use, and the perceived value of various social media as sources contributing to agenda building (Cobb, Elder 1983). If journalists are regularly monitoring sites and forums for story ideas and information, it is necessary for PR professionals within each industry to carefully monitor the information placed there and perhaps engage content producers. In this paper, I conclude that social media is an alternative instrument to encourage a two-way communication channel between government and publics. In public relations, the emergence of social media challenges the traditional instruments of government public relations. Responding to the development of information and communication technology (ICT), social media is considered as an alternative communication channel of government public relations efforts.

Keywords: social media, public relations, electronic PR, site, forum, PR models, Grunig, journalists.

1. WHAT IS SOCIAL MEDIA?

Social media has significantly altered the world of media unlike any other medium. With its viral, informal and unedited format, it will continue to grow and change and present new opportunities for both public relations and traditional journalists. The integration of social media into the overall media landscape represents a new challenge for PR professionals.

This new term is defined by many scholars as "a variety of new sources" (Mangold, Faulds 2009), "internet-based applications" (Blackshaw 2006), "a channel or a group of new kinds of online media" (Mayfield 2008).

Social media encompasses a wide range of online, word-of-mouth forums including blogs, company sponsored discussion boards and chat rooms, consumer-to-consumer e-mail, consumer product or service ratings websites and forums, Internet discussion boards and forums, moblogs (sites containing digital audio, images, movies, or photographs), and social networking websites, to name a few. (Mangold, Faulds 2009: 358).

1.1. Definitions

This form of media "describes a variety of new sources of online information that are created, initiated, circulated and used by consumers' intent on educating each other about products, brands, services, personalities, and issues" (Mangold, Faulds 2009: 358).

While there is a lack of a formal definition, 'social media' can be generally understood as Internet-based applications that carry consumer-generated content which encompasses "media impressions created by consumers, typically informed by relevant experience, and archived or shared online for easy access by other impressionable consumers"1 . According to various scholars, social media elements and mainly business blogs promote and humanize the two-way communication by enabling companies to talk with customers and allowing customers to talk back2 .

Social media not only gives the PR practitioners the opportunity of reaching out to and engaging their publics in conversation, but also provides an avenue to strengthen media relations. Brian Solis, the leading PR 2.0 "evangelist" and exponent, sums it up in his contribution to The Social Web Analytics eBook 2008 by Philip Sheldrake of Racepoint:

Social Media is no longer an option or debatable. It is critically important to all businesses, without prejudice. It represents a powerful, and additional, channel to first listen to customers, stakeholders, media, bloggers, peers, and other influencers, and in turn, build two-way paths of conversations to them. Yes, conversations are taking place about your company, product, and service, right now, with or without you. This represents priceless opportunities to build relationships and shape perceptions at every step. In the process, you become a resource to the very people looking for leadership, expertise, vision and solutions. The most important driver for outbound and proactive online relations is that it's measurable and absolutely tied to the bottom line. (Brown 2009: 18).

The real power of people can be noticed in the new revolutionized media channel, namely social media. According to Mayfield (2008), social media is best understood as "a group of new kinds of online media", which share most or all of the following characteristics3 :

1. participation and engagement: social media is a means of making everybody interested provide contributions and feedback. Thus the border between media and audience seems to disappear.

2. openness: participation imposes some freedom that the audience enjoys through voting, comments or sharing information.

3. conversation: unlike traditional media which is focused on one-way communication, social media provides a two-way conversation, thus communication is instantaneous.

4. community: the most important outcome of social media is forming communities. The backbone of these communities is a sharing of common interests.

5. connectedness: social media creates a sort of chain effect, making use of links to different sites, resources and persons.

Drury (2008: 274-277) suggests that when analyzing social media, marketers too often concentrates on the "media" factor, when "social" element is the key, because marketing within social media is about building a relationship and conversation with the audience, where the simple message delivery is changed by ongoing exchange of perceptions and ideas between company and the consumer.

Haven4 suggests that social media key elements are not entirely new as features of sharing, connecting, opining, broadcasting and creating has been long in our lives, but there are several characteristics of new technologies and behaviours that set them apart from the past:

1. reach – A tribe, a family, friends, neighbours or the local community have been social habitats which we belong to. Nowadays the new technologies offer a new opportunity of crossing the local borders in order to reach a global audience.

2. accessibility – The advantage of new technologies is to reduce the costs of traditional media and to allow freedom to access to everybody.

3. usability – One consequence of accessibility is the freedom provided to everybody to create and operate the means of production. Creative and technical skills are no longer compulsory.

4. transparency – Nowadays it has been observed a craving for information. The new technologies provide the space where everything can be posted by everybody. But this sort of transparency might fire back sometimes since there is little control on the content.

5. recency – Instantaneity seems to be the governing word nowadays. People become addicted to emails, facebook or twitter and they seem to have forgotten the pleasure of holding a letter in their own hands.

6. These features show that social media is an encompassing term which combines technology, social interaction, pictures, videos and audio materials.

1.2. Advantages of Social Media

In Optimizing Your Public Relations with Social Media5 , published on the site www.prnewsonline.com, some PR practitioners sustain that social media is important for a number of key reasons6 :

- stickier than traditional media – According to a study conducted by InsightExpress (a market research firm) consumer trust in advertising has decreased 41% over the past three years (Elkin 2005). The prototypical consumer of social media seems to enjoy the trust and the message retention provided by this means of communication. The traditional forms of marketing and advertising are downplayed by word-of-mouth which, according to WOMMA (Word of mouth Marketing Association), is favoured by 92% of consumers.

- viral nature – social networking provides an instantaneous passing of information which might be interpreted either as a substantial opportunity and threat for communications professionals.

- interactive – (Corporate) blogging is the means of getting immediate response from the publics. The possibility of enabling comments should be taken into account by organizations since they might gain valuable costumer feedback.

- high visibility on the net – Besides immediate communication, visibility is another keypoint that organizations should take into account. Wikipedia is perhaps the most notable example of this phenomenon. Spannerworks, a search engine marketing specialist, recently reported that social media platform Wikipedia appears in the top 20 Google search results for 88 percent of searches for the top 100 global brands (Mayfield, 2007).

1.3. Social media and Electronic Public Relations

Zerfab Ansgar (apud Vegheş Ruff, Grigore 2003: 26) says that:

(...) in terms of public relations, the key to online business success lies in building and maintaining reputation through the exchange of information between the organization and the public online (translated into English by – M.P.).

So, "public relations use the new interactive media which either allow an immediate and technical dialogue with different social groups (eg e-mail, newsgroups), or made available to a communication partner information in electronic form so that it can appeal to them directly (CD-ROM, WWW), or support the management process of public relations (eg.

Automating press reviews and media monitoring)". That is to say E-PR uses Social Media or E-PR means public relations on the Internet:

E-PR is an area of public relations which through the implementation and use of tools available from online media and methods and derived measures seek to establish relationships with relevant groups of an organization. This means that online public relations support the objectives of public relations in general, helping to balance the long-term relationship between an organization and its target groups. (Friedlander, apud Vegheş Ruff, Grigore 2003: 30; translated into English by – M.P.)

If so far the communication between an organization and its publics was through the media, the Internet now has to "enable an organization to become the sender in mass communication, the message sent via the Web being accessible anywhere and anyhow" (Vegheș Ruff, Grigore 2003: 31). Thanks to the social media, the communication is now transformed, so that the information arrives to the public through the Internet.

ORGANIZATION → MASS-MEDIA → PUBLICS ORGANIZATION→ INTERNET→ PUBLICS

PR through traditional methods	E-PR through social media
Transmission of press releases	Transmission of press releases Post press releases on sites
Dissemination of advertising lay-outs	Information on forums, blogs, websites, campaigns and so on
Press conferences	Video-conferences, press packs can be sent/posted as PDFs
Organization of events	Post pictures from events on sites, blogs and so on
Public communication by letter, fax, through the PR offices, audience	Public communicate by e-mail, finds out information directly from websites, and the audiences also take place online (eg. Ministry of Education)

As it can be observed, E-PR means sites, blogs, on-line information and communication mediated by Internet.

2. MODELS OF PUBLIC RELATIONS

An understanding of Grunig's Four Models of Public Relations which describe the evolving types of PR practice from Press Agentry through Public Information to Two-Way Asymmetric and Two-Way Symmetric communication is also important to a study of effectiveness of communication between public institutions and journalists through social media.

Grunig and Hunt's "4 Models" of public relations practice went on to become the most talked-about theory in the discipline. The "4 Models" describe different approaches to public relations in the context of a 130-year timeline that shows how public relations have evolved. In the process, Grunig and Hunt (1984: 21-22) identify an "ideal" approach to public relations, namely the 2-way symmetrical model — and place it at the top of the evolutionary pyramid.

The four models are:

- press agentry (one way communication; often media relations and sometimes of the spin/propaganda variety; nonconsultative)
- public information (one way communication; normally from the perspective of the organization only, so it can be onedimensional; non-consultative)
- two-way asymmetrical (stakeholder views are sought; communication is adapted to potentially change stakeholder behaviours; the organization does not change its views or behaviour; very much in the mould of modern marketing)
- two-way symmetrical (as the one-way symmetrical model, except the organization does change its views and/ or behaviour to meet its stakeholders 'half-way' [or thereabouts, at least]; characterized by negotiation and compromise, education).

The 2-way symmetrical model is based on a free exchange of information that is used to alter attitudes in both the organization and its publics. This model of public relations depends on good feedback and is used primarily by organizations that are governmentally regulated and must prove that they are socially responsible.

The four models of public relations, and especially the two-way symmetrical model, have been the most controversial and the most debated component of the excellence theory (Grunig, Grunig, Dozier 1992). Many authors (Yarbrough, Cameron, Sallot, McWilliams 1998) have essentially equated the entire Excellence theory with the two-way symmetrical model.

Organizations that apply the notion of the two-way symmetrical communication will:

- enjoy more mutually beneficial, sustainable and commercially satisfying relationships with their stakeholders;
- make a more significant and worthwhile contribution to society;
- have more committed employees who are stronger organizational advocates.

I consider that the application of two-way symmetrical communication is the best way to build and sustain long-lasting, mutually beneficial relationships between an organization and its stakeholders. It will also make a meaningful contribution to society by increasing understanding of various factors by often heterogeneous groups of people. Finally, and most importantly, as organizations change to meet their stakeholders' needs, society will become more satisfying and more equitable to greater numbers of people.

2.1. The advantages of communication through social media

If we take a short look at social media, we can say that its main advantages are (Vegheș Ruff, Grigore 2003: 32): constant communication, immediate response, global audience, knowing the audience, bidirectional communication, low cost.

Social media not only allows public relations practitioners to reach out to and engage their publics in conversation, but also provides an avenue to strengthen media relations (Eyrich, Padman, Sweetser 2008: 412-414).

So, the public sector needs to understand that there are three elements underlying their existence, namely:

- reputation: It is their social responsibility to embrace this change.
- perception: Their social media participation will play an essential role in how the public perceives them.
- measurement: Their achievement will be evaluated not just on what they did, but also on people's perception.

There is a chance happening in public engagement. From the position of a representative of public sector, you do not know how the result will be, but your engagement will determine how the outcome is going to be. Another aspect that should be taken into account is that you (public sector) are aware of the fact that you must be open to any changes. Social media is not just global, but also local, so it represents rich sources of local interest issues, news, stories and local communication channels. Social media is the means of a transformation of a world in which people knew where the barriers were laid into a world without limits. Each social media connection established on the digital landscape has the potential of adding (or reducing) meaning, understanding, life/ humanity.

Applications of bilateral symmetrical communication model define excellence in public relations, and thus the effectiveness of an organization. Giving all the characteristics of social media, we can say that they are tools that facilitate the implementation of bilateral symmetrical communication model favouring direct communication with all audiences, instant feedback and behaviour change.

3. METHODOLOGY AND RESEARCH DESIGN

Against this conceptual framework, this study focuses on the role of social media in government public relations. Journalists consider that the sites of the Romanian ministries are a source of information.

Hypotheses:

The PR officers of the Romanian ministries use social media as a means of information provider.

The journalists in Romania use this type of information in their articles.

Corpus: the 18 sites from the current government ministries and 8 journalists from different media7 .

3.1. The analysis of the 18 sites included the following sections:

News on homepage, Site updating, Media sections, Forum, newsletter, Scale of transparency: email webmaster/ questionsfor audiences other than mass-media/mail PR/Contact leader8 .

News on the homepage:

Of the 18 ministries, eight do not update their web sites daily: Ministry of Transport and Infrastructure, Ministry of Public Health, Ministry of Economy, Ministry of Finance, Ministry of Youth, Ministry of Tourism, Ministry of National Defence, Ministry of Labour.

Media Sections include:

Ministry of Culture	Archive Press releases 2005-present, Conference
Ministry of Economy	Archive Press Releases 2004-2007
Ministry of Finance	Archive Press Releases 2008-present
Ministry of Justice	Archive Press Releases 2007-present
Ministry of Labor	Archive Press Releases 2008-present
Ministry of Youth and Sport	Press releases 2009, archive photo 2009
Ministry of Tourism	Press releases 2009, news, Photo
Ministry of Foreign Affairs	Press releases, conferences, interviews
Ministry of National Defence	Press releases, conferences, interviews, courses
Ministry of Administration and Interior	Press releases, conferences, shows, news, archive, publications
Ministry of Agriculture, Forestry and Rural Development	Press releases 2009, information
Ministry of Communications and Information	Press releases, events, Photo gallery, Public information
Ministry of Regional Development	Press releases, information, invitations
Ministry of Education, Research	Press releases, Relations with the public
Ministry of Small and Medium Sized Enterprises	Archive press releases, Press conferences, Discourses, Photo gallery, Contact Press office
Ministry of Environment	Press releases, Events, Conference in a video format
Ministry of Health	Press releases, Discourses, Meetings
Ministry of Transport and Infrastructure	Press Releases, Invitations, Announcements, Photo Album, Video, Annual Reports

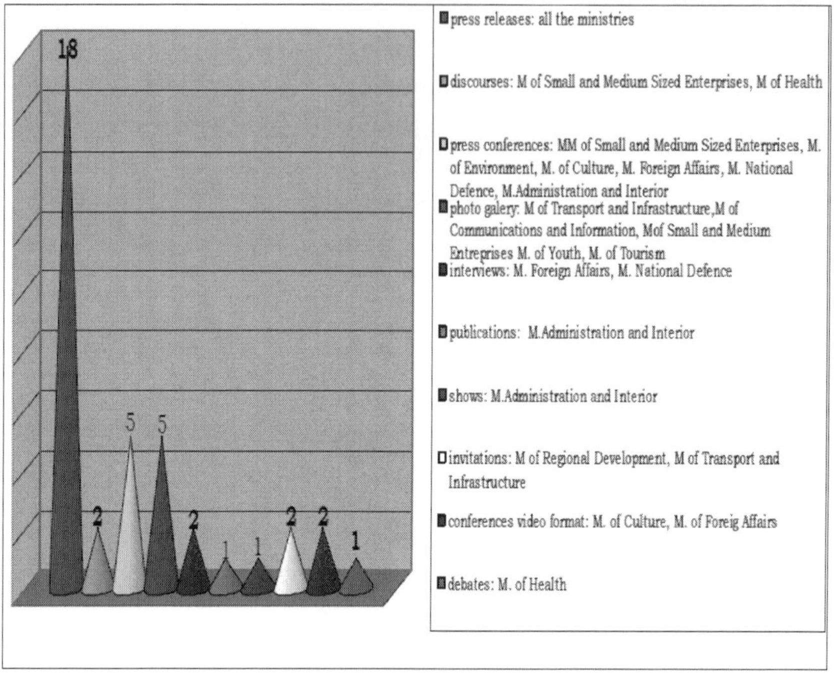

Figure 1. Types of PR materials made for the Media Section

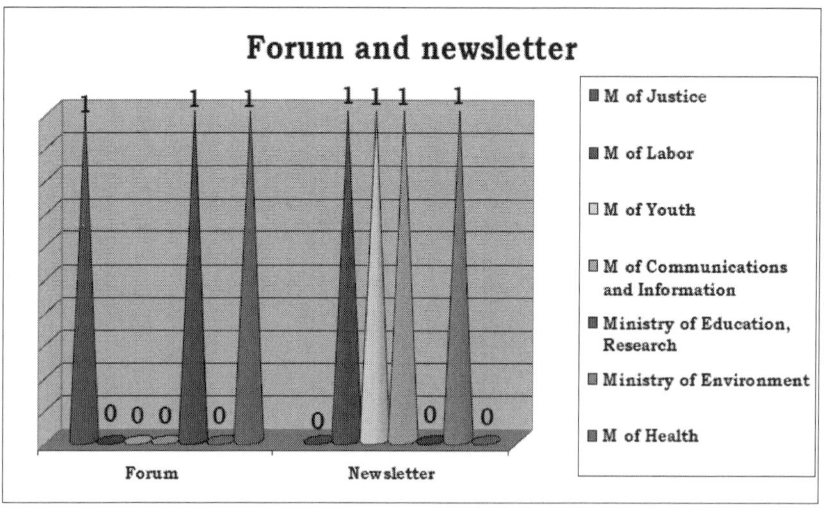

Figure 2. Forum, newsletter

Forums are the longest form of online social media. They most commonly exist around specific issues and interests, for example education9 .They are sites for exchanging ideas and information usually around special interests. Internet forums are places for seeking advice on a topic, for sharing news, for flirting, or simply for spending time with idle chat. In other words, their huge variety reflects the existence of implicit face-to-face conversations. These sites are moderated by an administrator, whose role is to remove unsuitable posts or spam. However, a moderator will not lead or guide the discussion.

Most people in public relations are probably well aware that e-newsletters can be much more powerful and useful than just automated alerts. An e-newsletter is a smart tool for reaching a number of constituencies, including the press, employees, potential investors, and prospective business partners. Clients with limited staff can make a single effort to develop a newsletter that meets the needs of the various constituencies. A good e-newsletter can bring more people to a client's Web site and expose them to the online pressroom and other offerings. However, this feature cuts both ways in that the newsletter must be carefully coordinated with the Web site so that newsletter recipients actually find what they are seeking on the Web site, but this cooperation can be difficult if the PR people are not responsible for Web site operations.

As we can observe, the e-newsletter is much more present than a forum.

This scale has been used by Curtin Patricia and Gaither Kenn to measure how open, or dialogic, a government web site is. This implies (...) the extent that organizations provide comprehensive information about their attributes and maintain timely

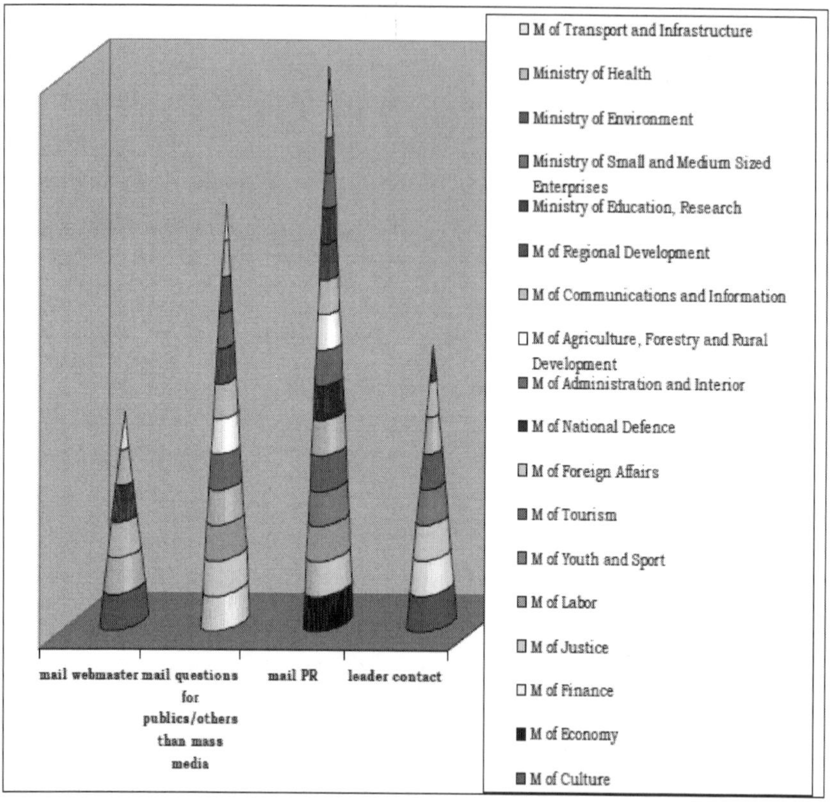

Figure 3. Scale of Transparency

communications directly to key public audiences. Transparency measures the effort made to make information available on the site and comprises three subscales: contact information, organizational and operational information, and freshness10 .

The most transparent institutions are the Ministry of Foreign Affairs and the Ministry of Communications and Information. The mail of the PR officer is more used as contact information than the mail for questions and the leaders' contacts.

Two sets of research questions were formulated, the former primarily aiming to understand the extent to which social media is used by journalists:

1. Do you access websites of ministries to search information?

All the journalists access the sites of Romanian ministries.

2. What sections do you visit?

AGERPRES – department of communication, PR and Press Office, org chart

ZIARUL FINANCIAR - news, press releases, Press office, organization contact

BUSINESS STANDARD - infrastructure projects, grant loans

RING - press releases, debates

ADEVARUL - press releases, org charts

ZIARE.COM -sections for press, draft legislation on public debate

RADIO ROMANIA ACTUALITATI - news, draft legislation

3. What type of information do you search in these sites? AGERPRES – press releases, legislative acts, the Minister's decrees, the Government's decrees

ZIARULFINANCIAR-statistics, implementing rules of the new decrees

BUSINESS STANDARD -infrastructure projects, grant loans

RING– debates, projects, statistics

ADEVARUL-press releases, org charts

ZIARE.COM–Ministries' projects, press releases, transcripts of Minsters' statements and speeches

RADIO ROMANIAACTUALITATI-press releases, projects

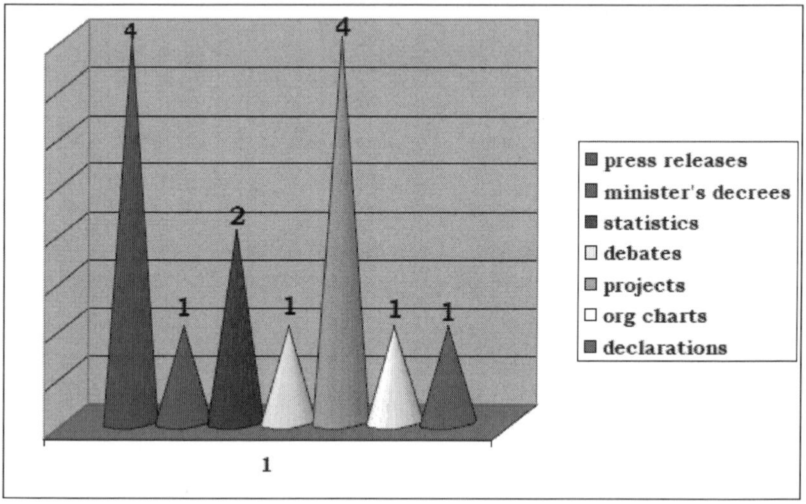

Figure 4.

Journalists usually search press releases and projects of the ministries, but statistics are also very important.

4. The sites contain sufficient information for editing or are only a source of information?

AGERPRES - For some articles, it is enough information on the site, but not for others.

ZIARUL FINANCIAR - Yes, they contain sufficient information.

BUSINESS STANDARD - Usually, the information, which interests me, could be very rarely found on the site. Most times, they are requested in writing and sent to the press office of the institution. Often, the answers need further explanations. Many of the members of the press office whom we ask questions do not provide sufficient information.

RING - Yes, they offer sufficient information.

ADEVARUL – Never do they contain sufficient information.

ZIARE.COM - Official communications are a primary source of information.

RADIO ROMANIA ACTUALITĂȚI – They do not contain enough data for editing; I always have to contact them by phone to receive more information.

Only two of the eight journalists consider that sites of Romanian ministries contain enough information for their articles. It is a key of fundamental journalism that reporters rely on two types of sources: primary and secondary. A primary source is one with firsthand knowledge of a story. Everyone else with knowledge relevant to the story is a secondary source. Stories are developed firstly around information received from the primary sources, which, in a perfect world, is further corroborated by other primary or secondary sources.

Every fact must be verified from a primary source. The primary source for a given fact is the source that originally generated that piece of information or the one that is able and authorized to report on that information firsthand. Common primary sources can include live experts, company literature, analyst reports, reference books, government agencies, and official organization websites.

5. Do you use the information on forums? Do you believe that information on the forums is true and that it should be checked?

AGERPRES - I have not checked if there are forums on these sites. Usually I do not use such information even if the sites of the Romanian ministries had such forums.

ZIARUL FINANCIAR - I read the forums, but there is no objective information.

BUSINESS STANDARD - Yes. But the information must be checked.

RING - I do not read forums.

ADEVARUL - I do not need to check the forums, especially that I do not have this possibility.

ZIARE.COM - The information on forums is trustworthy only if it is posted by the managers who run the forums. It must also be combined with other sources.

RADIO ROMANIA ACTUALITATI - I would not trust this information, it must be checked.

Information provided by the forums of the ministries must be verified by journalists because there are many articles which present forum opinions11 .

The second set of research questions was formulated to understand the extent to which social media is used by the PR officers from ministries:

1. What section of the website updates the institution for which you work?

2. What type of information is posted on the site and what sections are you interested in?

3. Who do you think the target audience of the materials on site is?

4. Do you watch the site traffic daily? Do you know which the most frequent visitors are?

5. Do you know the journalists who access your site in order to look for information? If so, what are the most commonly accessed sections?

6. How do you verify that the materials posted are journalistic source material on "the forum", "Press office/ communications with the media / information"?

7. If your site has a forum, who handles it and how often is it checked?

8. Do you consider that all the questions on the forum get a satisfactory answer?

9. Has it ever happened to write a press release or to develop a communication campaign, starting from a question on the forum? If yes, please exemplify.

From all the sites, www.edu.ro is monitored by trafic.ro. Some of the data collected are posted on trafic.ro website (usually www.edu.ro is first in the category Education and in the general ranking of the top 20 hits - Daily and overall).

Almost all sites of the ministeries are managed by a manager, other than the PR officer. Only the site of Ministry of Education and Research is managed by the Public Relations officers.

3.2. Results

Journalists consider that the sites of the Romanian ministries represent a source of information, the forums are helpful, but they do not contain reliable information. Communicators use social media provided by the websites of the institutions they work for. Of the 18 sites, only 3 are interactive, namely those of:

the Ministry of Education, the Ministry of Health and the Ministry of Justice. Regarding the information accessed and used by the media, the press releases and the projects are on the first place. The website of the Ministry of Communications and Information Society is the first website of a public institution in Romania to offer access to blind people: the user could read the speech content pages, the keyboard navigation being sufficient.

4. DISCUSSIONS

Journalists embrace the concept of social media more than they put it into practice. Non-interactive online information sources (web pages, directories) remain the important "first line" sources for journalists. While it appears that journalists do not use social media extensively in order to shape in an agenda-setting process, they are not against it. As such, and given the responses to the desire to work with the practitioners using social media, it is necessary for the PR officers to begin engaging social media within the process of agenda-setting. The most popular tool used by the PR officers from the Romanian ministries was the e-press release.

5. CONCLUSION AND FURTHER IMPLICATIONS OF THE STUDY

If public communication is a social binding tool (Zemor [1995] 2003: 37), then the citizen may be a client and

(...) the citizen wishes to be informed, to be taken into consideration, to be listened to. His criticism has a familiar character, meaning that – the more he loves, the more he punishes. The critical tendency is justified with the hope to see the procedures changed simplified (Zemor [1995] 2003: 37; translated into English by – M.P.).

Client status is not easy at all, but it may be more complicated, because a customer needs to be fully aware of his interests and to be able to decide for one of the offers provided.

Although there is the risk of using social media less, the practice of social media in government public relations has a positive feedback from citizens and mass media. If government public relations extend the fields of public service which social media handles and increase the level and scope of participation, government will better make use of the opportunities that social media might provide for government public relations. Government should address these challenges and it should take advantage of the utilities of social media in public relations.

Nowadays intensive observation of the PR practitioners' roles should indeed include the use of social network sites and social media tools.

ACKNOWLEDGEMENTS

The author would like to extend her gratitude to Roxana Moroşanu for her assistance on this project

6. REFERENCES

1. Brown, Rob (2009) Public Relations and the Social Web. London and Philadelphia: Kogan Page Limited.

2. Cobb, Roger W., Elder, Charles D. (1983) Participation in American Politics: The Dynamics of Agenda- Building. Baltimore: John Hopkins University Press.

3. Drury, G. (2008) "Opinion piece: Social media: Should marketers engage and how can it be done effectively?" [In:] Journal of Direct, Data and Digital Marketing Practice. 9 (3); 274-277.

4. Eyrich, Nina, Padman, Monica L., Sweetser, Kaye D. (2008) "PR practitioners' use of social media tools and communication Technology." [In:] Public Relations Review. 34; 412-414.

5. Grunig, James E., Grunig, Larisa A., Dozier, David M. (1992) Excellence in Public Relations and Communication Management. Hillsdale NY: Lawrence Erlbaum Associates.

6. Hunt, Todd, Grunig, James E. (1984) Public Relations Techniques. Philadelphia: Holt, Rinehart and Wiston.

7. Mangold, W. Glynn, Faulds, David J. (2009). "Social media: The new hybrid element of the promotion mix." [In:] Business Horizons. 52(4); 357- 365.

8. Theaker, Alison (2002) The Public Relations Handbook. London, New York: Routledge.

9. Vegheş Ruff, Iulian, Grigore, Bogdan (2003) Relaţiile publice şi publicitatea online. Iași: Polirom.

10. Zemor, Pierre ([1995] 2003) Comunicarea publică (La Communication Publique, translated by Margareta Samoilă, Ion I. Ionescu). Iaşi: Institutul European.

7

CHAPTER

A General Outlook on Turkish Librarianship and Libraries

Hilmi Çelik, Didar Bayır

Director, Information Centre, Sabancı University, Orhanlı, 34956 Tuzla, I' stanbul-Turkey,

Director, Suna Kıraç Library, Koç University, Rumelifeneri Yolu, 34450 Sarıyer, I' stanbul-Turkey,

ABSTRACT

Turkey has a rich book and library tradition going back centuries. This article presents an overview of Turkish libraries and librarianship by providing information specifically on the library tradition in Turkey, public and special libraries, librarianship education, professional associations and publications, and recent developments including some special projects run by the state and NGOs.

Keywords: Turkish libraries, Turkish librarianship, Turkish librarianship education, Turkish librarians' associations

GENERAL FRAMEWORK

Turkey has a rich book and library tradition going back centuries. This tradition has continued to these days as the Ottoman sultans and other wealthy people supported the production of manuscripts and established libraries. In 1729 the first printing house was established, and after the 1850s libraries and their collections started getting better organised within governmental institutions. The period between 1856 and 1876 is notable for the pronounced progress in the field of traditional Turkish librarianship. The Beyazit State Library, the first library established by the state in 1882, is the result of this progress, as was the publication of the first printed library catalogues in 1884–1886.

One of the greatest reforms realised by the Government of the Republic is the Educational Reform of 1924. With this reform, the management of all waqf libraries, established for different purposes over the years and in a sense serving as public libraries, was entrusted to the General Directorate of Libraries operating within the Ministry of Education. Today, the General Directorate of Libraries and Publications, as it is now called, and which operates under the Ministry of Culture and Tourism, is responsible for all the public and manuscript libraries in Turkey.

By means of the Legal Deposit Law, which was enacted in 1934 and still is in force today, an effort was made to organise nationwide supervision of all printed material. By law it was made mandatory that all printing houses supply six copies of the material they printed to the Directorate of Depository. Beyazit State Library, Istanbul University Central Library, Ankara National Library, the Library of Parliament, Adnan Ötüken Provincial Public Library and Izmir National Library are the libraries benefiting from this law. Work is now underway to adapt the law to the present-day requirements.

The idea of 'establishing a national library' had been around since 1871. All attempts at this, however, were fruitless until 1948, when new efforts bore fruit and soon after, in 1950, the Turkish National Library was established by a specific law that was the first of its kind in this field.

At the same time so-called 'community centres' (halkevi) and village community centres (halk odasi) which had been established by Atatürk on 19 February 1932 grew into public libraries and culture centres. There were over 500 such centres in 1936. However, due to various political reasons these community centres were closed in 1951 and segments of their collections were incorporated into libraries affiliated with the Ministry of Education's General Directorate of Libraries.

Public libraries which started off as waqf (vakif) libraries and continued as community centres, and which today also include children's libraries, are continuing their activities. Of late plans have been developed to turn over public libraries to municipal administrations.

According to 2007 statistics, there are 1,163 public libraries in 81 provinces. These libraries house 13,198,814 publications, and 2,796 staff annually serve 20,228,517 users. Sadly, only 333 of these staff members are librarians. In addition, library services are being provided by 59 mobile libraries to 763 areas nationwide.

Turkey continues to work on expanding public library services on a national scale. In addition, Turkey participates in collaborations on an international level it is involved in exchanging knowledge and skills. In this context, between 2002–2003, Turkey participated in the PULMAN-XT () Project which was developed by EU member and candidate states with the aim of strengthening the potential roles of public libraries, museums and archives towards e-Europe. As a result, a significant step was taken towards restructuring public libraries, promoting best practices and increasing their numbers. Subsequently, Turkey participated in yet another European Union Project called CALIMERA () which paved the way for promoting cooperation between public libraries and other local cultural institutions (archives and museums).

Other projects spearheaded by certain non-governmental organisations and some individuals and institutions followed this significant work towards the modernisation of public libraries. Significant nationwide examples of such projects include the 'renaissance in the Villages' project through Rasime – Recai şeyhoğlu's village libraries, the 'My Library' project by the Community Volunteers Foundation in coordination with university students and the 'For Libraries' project started by the Ministry of Culture and Tourism, which calls for for support in many ways, ranging from construction of library buildings to donation of books.

Various Turkish works are being published in different languages within the scope of the TEDA Project, an effort in translation and publication support started in 2005 by the General Directorate of Libraries and Publications of the Ministry of Culture and Tourism, with the purpose of 'publicising Turkish culture, art and literature overseas and the languages of the world to gain from the riches of the Turkish literary language'.

Services provided to disabled citizens at public libraries have been expanded. Talking books were produced, specifically for the blind, and CD recordings were made available for checking out. Within the scope of the 'Talking Books' Project of the Ministry of Culture and Tourism, books on various subjects from science and technology to literature were recorded and made available on the website of the Ministry for the use of visually impaired citizens.

In 1997, automation was introduced and preparations started for a project to transfer the catalogue records of the collections at the public libraries affiliated with the General Directorate of Libraries and Publications to a central database and then be made accessible to the public. In 2006 the project was finalised and MİLAS Library Automation Software was made available to libraries for shared use. At the moment 385 libraries participate in this system, and 3,633,000 bibliographic records have been transferred onto electronic medium by April 2008. These records are all available from the main server. Libraries with the appropriate infrastructure (meaning those equipped with at least one computer and an internet connection) will be added to the system as soon as possible. The main purpose is to promote the widespread use of a single interface system in all our libraries, thereby push usage levels to the highest level.

Within the 'Public Libraries Internet Access Centres Project', members at 81 provincial public libraries and Istanbul Beyazit State Library are being given free internet access since 2006. The aim is to increase the number of libraries providing internet access to 400 in 2008. The 'Books Delivered to Home Project' was started in 20 pilot provinces for home-bound, ill and disabled citizens who cannot make use of established libraries. The General Directorate of Libraries and Publications became a partner in the READCOM project financed by the European Union within the framework of the Grundtvig 2 Adult Education Programme, and as a consequence reading clubs for adults were formed.

Aside from public libraries affiliated with the General Directorate of Libraries and Publications of the Ministry of Culture and Tourism, there exist public libraries established by municipal administrations. The best and established example of such a library is the Atatürk Library in Istanbul, which houses numerous historical resources and as such has the identity of a research library.

In 2007 Turkey was introduced to the 'Boutique Library' idea. The boutique library opened by the Municipality of Kadiköy in Istanbul became the focus of attention with its up-to-date arts oriented collection and technological hardware.

Turkey has a rich diversity of private libraries and there are specialised libraries alongside the manuscript libraries, and local and foreign cultural centres' libraries, e.g., the Women's Library and Information Centre Foundation established in 1990. The libraries of the German Archaeological Institute, the Orient Institute, the French Institute for Anatolian Studies (IFEA), the Netherlands Institute, the Swedish

Research Institute and the American Research Institute (ARIT) can be mentioned as examples of foreign research libraries. In addition, a number of cultural centres such as the Goethe-Institut, the Institut Français, the Istituto Italiano di Cultura, the Instituto Cervantes and the American Information Resource Centre have their own libraries. Aside from providing information and documents on their own cultures, these centres contribute to the cultural and social life of the city in which they are located.

The Borusan Music Library with approximately 8,500 books, 10,500 CDs and around 7,000 musical scores and the Yapı Kredi Sermet Çifter Research Library, with around 80,000 manuscripts and printed books, are just two examples of the domestic cultural centre libraries. The library of the Research Centre for Islamic History, Art and Culture (IRCICA), a subsidiary organ of the Organisation of the Islamic Conference (OIC), located within the Yıldız Palace complex in Istanbul, with its extensive collection is also worth mentioning.

Today, there are over 50,000 schools at the elementary, secondary and vocational/technical levels in Turkey. These schools are supervised by the Ministry of Education as are their libraries and, sadly, it cannot be said that the libraries of these schools are sufficiently developed. However, in some schools, most of which are private schools, it is possible to find libraries that are considerably better organised. Even though the 'School Libraries Regulation' of 1950 (and updated in 2006) required these schools to establish libraries, it has not been possible to realise their establishment, development and management, nor their management by trained professionals. Their collections were not kept up to date and do not complement educational programmes. Nonetheless, there are some private school libraries with vocationally trained librarians and purposive collections located within easy reach of students. Some of these schools play important roles in contemporary schooling/educational programmes. Their collections can comprise between 3,000 and 50,000 volumes. School librarians teach subjects such as information literacy, proper use of resources, internet security and cooperation with the teachers which complement school curricula. Although few in number, some school libraries have e-catalogues, computers with internet access and different databases, and with these resources are directly contributing to education.

In general it can be said that the need for public and school libraries has increased, and issues such as the need for organising and technically facilitating access to information and the need for creating buildings which conform to international standards, have found their way to the country's agenda.

EDUCATION OF INFORMATION PROFESSIONALS

Four Turkish universities have a Department of Information and Document Management which educates librarians, archivists and information specialists at the undergraduate level and also offers graduate degrees (Ankara University, Istanbul University, Hacettepe University and Marmara University). The aim of these departments is to educate both personnel who can generally undertake information and document management successfully as 'information professionals' and personnel who will be researchers in the field of information-document management for the public and private sectors. The first such Department was established at Ankara University in 1954–1955 with the support of the Ford Foundation and the American Library Association. The second Department of Librarianship in Turkey was opened at Istanbul University through the efforts of Rudolf Juchhof, a German Professor. Hacettepe University's Department of Librarianship was set up in the 1972–1973 academic year, at first only at the graduate level and later with BA-level education. Marmara University, with its Department of Archives, opened up in 1987 and having been renamed Department of Information and Document Management in 2002, it has taken its place among those institutions educating information professionals.

PROFESSIONAL ASSOCIATIONS AND SUPPORT GROUPS

There are two non-governmental organisations acting as professional associations for the librarians in Turkey:

The Turkish Librarians' Association (TKD): The Turkish Librarians' Association is the first vocational organisation in Turkey. Since 1949, it has been actively serving as a non-governmental institution of the world of books and libraries and as the voice of librarians. Today, it has 25 branches and 1931 members. From 1952 onwards, it has been publishing the periodical Turkish Librarianship. There have been 154 book publications. Each year, TKD, under the auspices of the Ministry of Culture and Tourism, and with the cooperation of Directorates of Provincial Culture and Tourism and National Education and all other libraries, organises the Library Week during the last week of March. During this week professional issues and expectations are discussed and social activities organised. In addition, TKD organises on-the-job training programmes with the purpose of assisting colleagues' integration into the contemporary world of information. TKD cooperates with national and international professional organisations; it is a member of IFLA and LIBER and it played an important role in bringing the 1995 IFLA Annual Conference to Istanbul. In addition, TKD participates in book fairs, it puts on exhibitions, and it organises book donations and activities such as trips and dinners, aimed at increasing social solidarity.

The University and Research Librarians' Association (ÜNAK): The University and Research Librarians' Association was established in 1991 with the purpose of examining issues arising from the rapid change and advance of technology concerning university and research libraries, documentation and information centres and private libraries, and comparing these to similar issues and solutions in other countries; investigating and resolving different issues concerning the personnel working in these information centres; implementing the application of new technology. In short, to realise what is required to get these information centres to be compatible with the contemporary librarianship concept and operations. At present, ÜNAK has 630 members and itself is a member of international organisations such as LIBER. ÜNAK, along with its standard professional and social activities, has been publishing the periodical Information Worldsince 2000 and it has been involved in a consortium effort together with international organisations such as OCLC.

The Innovative Library Initiatives Promotion Group (ILIPG): The Innovative Library Initiatives Promotion Group is made up of a group of librarians carrying out and supporting various activities towards goals listed below:

1. to accelerate the upgrading of public libraries in Turkey,
2. to activate and/or develop the librarian and global network, on an international level and in Turkey,
3. to develop lobbying activities for libraries in Turkey,
4. to develop professional and technical abilities of librarians' and librarianship students.

For this purpose, ILIPG organises regional and national workshops with local and foreign participants, it provides short-term work experience in 'model' libraries locally and overseas, it translates documents on innovative developments/new standards from English or other European languages (German, French, Spanish, etc.) into Turkish. ILIPG works in close cooperation with the Turkish Librarians' Association

(TKD). It is made up of librarians from various libraries in Istanbul. These include librarians from university libraries and foreign culture centres' libraries together with faculty members from Departments of Information and Document Management

PROFESSIONAL PUBLICATIONS

Publications concerning librarianship are generally realised by professional associations and Information and Document Management departments.

However, it is hard to say that these publications are sufficient. Nevertheless, two professional periodicals have to be mentioned:

- Turkish Librarianship (Türk Kütüphanec-iliği) – A peer-reviewed journal published every 3 months by the Turkish Librarians' Association (TKD). This journal, included in the 30 oldest journals

published according to world standards, took its place on the international platform in 1994 and was indexed in the LISA (Library and Information Science Abstracts) and electronically published in 1999.

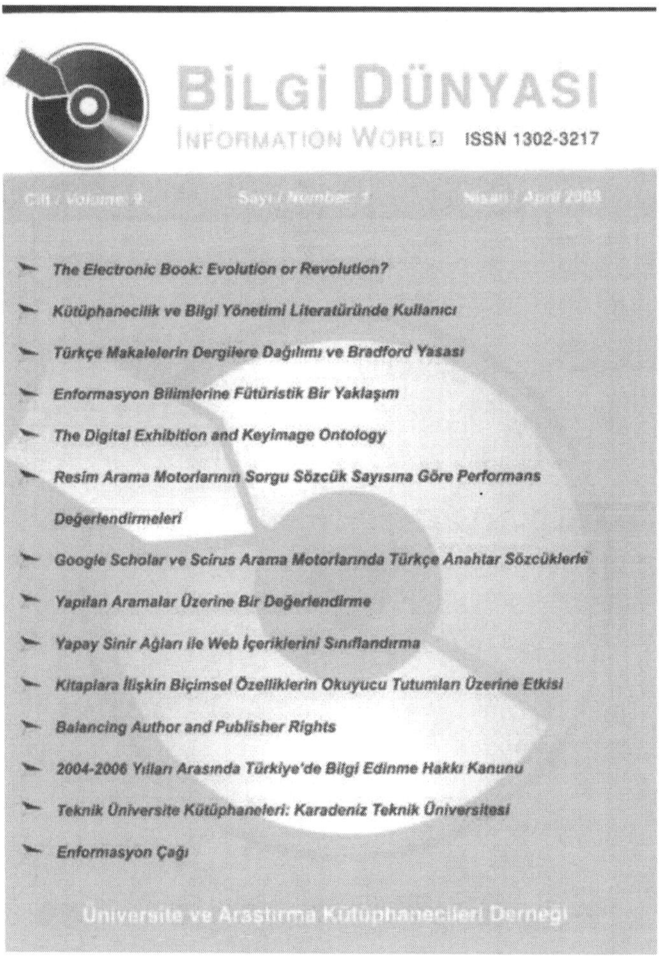
BİLGİ DÜNYASI
INFORMATION WORLD ISSN 1302-3217

Cilt / Volume: 9 Sayı / Number: 1 Nisan / April 2008

- *The Electronic Book: Evolution or Revolution?*
- *Kütüphanecilik ve Bilgi Yönetimi Literatüründe Kullanıcı*
- *Türkçe Makalelerin Dergilere Dağılımı ve Bradford Yasası*
- *Enformasyon Bilimlerine Fütüristik Bir Yaklaşım*
- *The Digital Exhibition and Keyimage Ontology*
- *Resim Arama Motorlarının Sorgu Sözcük Sayısına Göre Performans Değerlendirmeleri*
- *Google Scholar ve Scirus Arama Motorlarında Türkçe Anahtar Sözcüklerle*
- *Yapılan Aramalar Üzerine Bir Değerlendirme*
- *Yapay Sinir Ağları ile Web İçeriklerini Sınıflandırma*
- *Kitaplara İlişkin Biçimsel Özelliklerin Okuyucu Tutumları Üzerine Etkisi*
- *Balancing Author and Publisher Rights*
- *2004-2006 Yılları Arasında Türkiye'de Bilgi Edinme Hakkı Kanunu*
- *Teknik Üniversite Kütüphaneleri: Karadeniz Teknik Üniversitesi*
- *Enformasyon Çağı*

Üniversite ve Araştırma Kütüphanecileri Derneği

- Information World (Bilgi Dünyası) – A peer-reviewed journal published every 6 months by the University and Research Librarians' Association (ÜNAK). It is indexed in LISA (Library and Information Science Abstracts) and EBSCOHOST-LISTA (Library, Information Science & Technology Abstracts).

ACKNOWLEDGEMENTS

Translated by Ali Orbay.

REFERENCES

1. Akbulut, Mustafa and Atılgan, Dog˘ an, 'Libraries and Librarianship in Turkey', *IFLA Journal* 21(1995), p. 83–88.

2. S, enalp, Leman, 'The Place of Libraries in the Atatürk Period Education System', *Atatürk Research Centre Journal*, VII(1991)21.

3. Yılmaz, Bülent and Bayır, Didar, 'PULMAN-XT Project and Its Effects on Turkish Librarianship', *Türk Kütüphanecilig˘ i* 18(2004)1, p. 79–84.

4. General Directorate of Libraries and Publications website (http://kygm.kulturturizm.gov.tr/BelgeGoster.aspx?F6E1 0F8892433CFF0D262A49C727F232C312D1DD2E9EA986) - accessed 10 February 2008

5. TÜI˙ K, Turkish Statistical Institute, http:// www.turkstat.gov.tr

8

CHAPTER

Towards a Better Design: Physical Interior Environments of Public Libraries in Peninsular Malaysia

Suhaila Sufar[a],, Anuar Talib[b], Haris Hambali[c]

[a] *Department of Built Environment (AP780) Faculty of Architecture, Planning & Surveying, UiTM Shah Alam*

[b] *Department of Interior Architecture, Faculty of Architecture, Planning & Surveying, UiTM Shah Alam*

[c] *Department of Interior Architecture, Faculty of Architecture, Planning & Surveying, UiTM Perak*

ABSTRACT

The paper will focus on how to create a better design of physical interior environments in public libraries in Malaysia. The research will consolidate on the interior ambiences caused by selection of lightings, furniture, materials and finishes of public libraries. It will discuss on the previous studies and literature reviews which aimed to evaluate and review their opinions on the influence of physical interior environments on users' expectations, needs and behaviors on the interior settings. This overview will help the researcher in future study to investigate on the interplay between library users and physical environment, particularly on spatial ambience and the users' behaviors.

1. INTRODUCTION

The public library is an importance as community assets. The subject that is being endlessly debated is about the future of libraries. Nowadays it is a place of knowledge acquisition, communication and socializing with other people. Public libraries and university libraries start to change the "rusty" approach of a quiet place to a new modern place of meeting other people and using information in groups. We can also see how the architecture and interior in libraries has changed during the time. New shapes, colors and concepts are employed in newest projects and interior solutions in different libraries all around the world. According to Juhnevica & Udre (2010), the needs of users, interior space planning and physical interior environment design should be at the centre of the planning process. Future of libraries should serve multiple roles communities and society. However, the success of the design will be achieved by the user's satisfactions on how the physical interior environments influence them. Latimer, K. (2007), points out that, user space needs to be well planned, welcoming and attractive as the role of the physical library and the needs of users change in the 21st century. The importance and interest of the study is to show that such issues as interior design and architectural solutions can be important in creating appropriate and friendly place for library users.

2. BACKGROUND

According to UNESCO (1999), shows that the report of reading habit among Malaysians were only read a couple of books a year. This report shows that, we are far left behind contrast with other developing countries. This obviously shows that the reading habits among Malaysians was still low and need to be generated with more consolidation for improving and in producing knowledgeable society and community. When it comes to the public library scenario in Malaysia, it is very different and far out to libraries in developed countries. This includes the interior design quality, comfort level, satisfaction, customer experience and the characteristics of the physical design of the interior environment. In researcher opinions, now is the time to address the issues in lack of reading habit among Malaysian and to recognize the reason why Malaysian people not interested to visit libraries. It may be related to less attraction to visit, lack of facilities infrastructure and less comfortable in interior space.

UNESCO's statistics (2009) also placed Malaysia at 11th (of 137 nations) that allocates a huge budget for national education. In the effort to create a society with knowledge, not only formal education comes into focus, instead inculcating reading habits is among the core values of a society with knowledge. Here, the role of the traditional libraries as bookstores of information, a quiet and bored place to visit, now becomes a place of acquiring knowledge: through socialization, communication, in a relaxed interior space. Despite the developments of accessible information and the "online" technology today, the physical library is still important as a public building that facilitates the acquisition of knowledge.

In some countries the library may be the only physical space for learning and sharing information in a non-institutional context. Other societies offer a wide range of possibilities and it becomes crucial for libraries to deal with different partners in lifelong learning and education. High-tech learning centres have the potential to become attractive working environments for the internet generation when they integrate technology in an appealing overall picture with communal and social spaces. The open learning space at last can serve the knowledge societies by offering low intensive and collaborative meeting places for their communities. The ideal way would be a comprehensive, multifunctional space combining all three approaches beneath the same roof. A library built and organized with this in mind could become an attractive place for informational participation and lifelong self-paced learning. (Eigenbrodt,2009)

Furthermore, the development of library building concepts should be evolutionary with new inspiring design and features appearing as the changing needs of the people. Library design is not only about the exterior envelope but is also about practical and exciting physical interior spaces and environments. One of the most important aspects for designing a library is to consider the physical interior environments which comprises of interior space planning, and interior ambience such as selection of lightings, furniture, materials and finishes.

The research questions of this research are:

- What is the physical interior environment should be considered in designing public libraries?
- What are user's expectations on physical interior environments in public libraries?
- How well the physical aspects and design of the public library had catered to the needs and expectation of users?
- What are the key criteria to achieve a better design in physical interior environments in public libraries of today and future?

The objectives of the research are:

- To study and explore on the various functions and related of physical interior environments in public libraries.
- To study the factors that related to the user's expectation on physical interior environments in public libraries.

- To study on user's perception and behaviors towards the physical settings of interior environments in libraries

Significance of the study is:

Firstly this study evaluates the physical interior environments through expert views, user's perception and their needs. Secondly, the study will improving the existing guideline and information which are produced before by the National Public Library, Malaysia (NPN) and Malaysia Public Works Department (JKR) in order to establish a key criteria of physical interior environment to achieve a better design of public libraries in the future. The study also will be based on evidence base design; which study on user's behaviors towards the ambience of interior environments.

3. LITERATURE REVIEW

This paper will begin with a review of the recent literatures in the field of library building design. The discussions will be consolidate on the definitions of public library, what are the criteria of physical environments which regards to the interior ambience such as various functions and importance of physical interior environments in public library, and the impact of the interior ambience and physical interior environments which influence human behaviors and user's experience.

3.1. Definition of Public Library

Generally, library is the central sources of knowledge and information have the same basic objective of providing services to disseminate knowledge and provide reference materials to customers who need the service. Libraries in the digital age of today, not only physically as a space for knowledge, but also a place for individuals or groups come together to explore, learn, meet, interact, socialize as well as enjoy democratic access to resources and information. Therefore, providing a comfortable space, meet the taste and satisfaction of interior space and physical environment in general is very important to attract users to come and use the library services.

According to UNESCO (United Nations Educational, Scientific and Cultural Organization), a public library is defined as the local centre of information, making all kinds of knowledge and information readily available to its users. The services of the public library are provided on the basis of equality of access for all, regardless of age, race, sex, religion, nationality, language or social status. Specific services and materials must be provided for those who cannot, for whatever reason, use the regular services and materials, for example linguistic minorities, people with disabilities or people in hospital or prison. All age groups must find material relevant to their needs. Collections and services have to include all types of appropriate media and modern technologies as well as traditional materials. High quality and relevance to local needs and conditions are fundamental. Material must reflect current trends and the evolution of society, as well as the memory of human endeavor and imagination. Collections and services should not be subject to any form of ideological, political or religious censorship, nor commercial pressure.

According to IFLA (Institute Federal Library Association, (2007), a public library is an organization established, supported and funded by the community, either through local, regional or national government or through some other form of community organization. It provides access to knowledge, information and works of the imagination through a range of resources and services and is equally available to all members of the community regardless of race, nationality, age, gender, religion, language, disability, economic and employment status and educational attainment.

McCabe & Kennedy (2003), add that the public library is a place that belongs to everyone. New libraries often are arranged like bookstores, and many offer light refreshments that encourage customers to linger. Libraries are increasingly becoming places for interaction, and most new ones have meeting spaces, seminar rooms, and sometimes performing- arts spaces and galleries. All of these activities and attractions make the library a desired destination and increase the square footage needs for new libraries.

According to Rooney-Browne (2009), she points out that the public library spaces should transform into vibrant, welcoming destinations; exploiting new revenue streams;attempting to establish spaces in virtual communities; and with the help of a SWOT analysis. SWOT analysis is refer to S-strenght, Wweakness, O-opportunity, T-threat.

3.2. Physical interior environment

Physical interior environment are one of the major issue that have been continuously debated and discussed in order to create a better design and attraction to the users. McCabe & Kennedy (2003), point out that, a library building must be attractive and aesthetically pleasing to the eye. Internally, it must be functional and current in the use of technology supporting services to its community. Architectural design features provide for the physical appearance. The descriptions that follow will appear to meld issues of attractiveness and effectiveness, because sometimes they are inseparable. The descriptions are applicable both to new buildings and to buildings requiring renovation.

The physical environment encompasses both architectural elements such as physical layout, furniture, and equipment and visual sensory elements such as color, texture, and lighting. These two aspects, in conjunction with ambient factors, create the interior environment. (Bitner M. J., 1990). In an upscale restaurant, for example, patrons expect well-prepared food and attentive service, but they also expect comfortable seating, mood-setting lighting, pleasant or luxurious décor, soothing music, and an opportunity for social interaction. In these situations, the emotional components of the service setting become more important and can strongly influence the consumer's ultimate assessment of the quality of the service as a whole. (Jiang, 2006).

According to (Anandasivam & Cheong, 2008), the biggest challenge is to gain more teenagers and young adults to the physical library. The library needs to become "cool" and comfortable so that teenagers could consider it as a good place for meetings and hanging out together. Young people prefer comfortable workplaces and more freedom to move around and explore the space; they need a place to use their laptops and different zones to work. Many young people prefer to work in open space workplace together with others but still some of them want to work in silent rooms. There is also a big need of silent rooms in the library, cited in (Juhnevica & Udre, 2010).

Furthermore, the libraries need to have all technological equipment to make working in the library as easy as possible (Childs, 2006).Users need to have approach to printers, scanners and electronic catalogues. Also the planning of the space in the library is very important so that users would not feel lost. The best approach is to plan all storeys in the same way. There should be different zones for different purposes – reading, web browsing, group works, references and recreation. There are many interesting ways how to make library building more comfortable and cozy, (Juhnevica & Udre, 2010).

3.3. Physical interior environment criteria's and parameters

In designing a better way of physical interior environments in public libraries, there are several criteria and parameters that need to be taken into account. The physical environment comprises of elements of interior such as physical layout, selection of furniture, and equipment and visual sensory elements such as color, texture, materials, finishes and lighting. These two aspects, in concurrence with ambient factors and creating the interior environment. On the other hand there are modern libraries which have a lot of open space, nice and comfortable furniture like sofas and armchairs, coffee-tables and many computerized places or workspaces for users to work with their laptops. Most of the libraries have free Wi-Fi Internet access, electronic catalogues and other technical equipment to make library functions more available for a greater range of users. It all matters when we look at latest trends in library interior and design. (Juhnevica & Udre, 2010). All senses can be involved – music, nice but not annoying smell, good lightening, comfortable furniture, pictures on the walls and many other things can make users feel welcomed. Variety of different furniture, different zones and places to gather or be alone can provide different types of users with just what they need. The library might become a very good place for exhibiting art and at the same time it could create a special atmosphere and make the library more dynamic (Anandasivam & Cheong, 2008).

3.3.1. Color selection

To begin, as McCabe, G. B. (2003) puts it, should study on the psychological effects of color. Dark colors may subdue excitable behavior, bright color will stimulate. According to Gold Coast City Council Branch Libraries (GCCC), 2007, a contemporary approach to color and materials selection should be adopted so as to evoke an inviting and friendly image with a sense of efficiency, coupled with a stimulating memorable building experience and civic presence. The new library building should demonstrate a clean line aesthetic, backed up by a layered, textural and timeless approach to the colour and materials palate. Colour should be incorporated selectively and based on a neutral 'background'. Colour may feature on elements such as selected walls and partitions, the fabric selection for furniture and the use of graphic elements to clearly define distinct areas of the library. Colours and finishes should be appropriately selected to reinforce the appeal of the spaces to the designated user age group, while seamlessly meshing with the total concept of the space as an inviting, stimulating place to visit. Nowadays there are more bright colours in the libraries.

3.3.2. Furniture and equipment

Variety of comfortable furniture and different furniture, different zones and places to gather or be alone can provide different types of users with just what they need. (Anandasivam & Cheong, 2008).

Standard Shelving -The consultant and the librarian will calculate the need in linear feet for bound volumes with an allowance for growth. This figure will be converted into shelving units, single or double faced, and in heights ranging from 42", or about 60" to 66", to full 84" and 90" heights. The shelving preferably should be steel. Some librarians may prefer the lower height shelving choices with very few units, if any, above the 66" height. The preferred unit has the base on the floor with shelves above. T-based shelving isn't recommended for two reasons: losing the lower shelf and the difficulty in cleaning under them. If the ethnic make-upof the area population tends to be people of shorter stature, then the lower Furnishing and Equipping the Library and Its Environs 45 shelving may prove more satisfactory. Capacity in volumes will be lower because of fewer shelves. (McCabe G. B., 2003). According to (Gold Coast City Council Branch Libraries (GCCC), 2007), The maximum height of book-stacks is limited to 1500mm above the finished floor to reinforce the fit-out transparency and to be in line with OH&S and Equitable Access guidelines for staff and customer use.

Seating- Seating space is a crucial element in library design. Most public and academic libraries find that space to seat the growing user population is increasingly encroached upon by shelving to house growing collection. Library users should have a variety of seating to choose from to match their learning approach and mood at a particular given time there are essentially three types of seating: 1) Reading benches,2) Reading tables,

3) Flexible seating. Based on the user feedback on seating, they find it convenient to form informal discussion spaces with the movable seats. Students are attracted to this area by the striking color and "cool" atmosphere. (Anandasivam & Cheong, 2008)

3.3.3. Lighting

Lighting can control how the library looks and feels. Good lighting is needed for study, but intimate lighting may be excellent for reflective areas. Use spot lighting for special effects. Planners of both new academic and public library buildings are also concerned with the external appearance of their facilities. In his paper on Flushing Library in New York, Gary Strong praises the curtain wall façade of the building which "allows the public to see library activity within and invites them into the building" (Strong,2001:127). This concept of transparency remains a popular feature in libraries today, such as the Lanchester Library

(Noon, 2004) and the Idea Store in Tower Hamlets, where "large amounts of glazing to enable local people to see what was going on inside were essential" (Wills, 2004:109). At the Hatfield Campus Learning Resource Centre the designers took care to build the glass façade along the northern side so as to reduce solar glare problems for the readers inside (Martin, 2004). However, McCabe (2000) outlines the various issues that arise from using a large amount of exterior windows in a library, including security and safety as well as avoiding glare and overheating."The earth has its limits, and buildings need to be environmentally friendly as well as handsome and cost-efficient." (Sannwald, 2007:135) The IC (University of Sheffield, 2007a) is of course one example of this – another is the Lanchester Library (Noon, 2004) which uses lightwells and natural ventilation to reduce the need for artificial light and air conditioning, and has windows carefully aligned to minimize solar gain and glare.

(Pierce 1980, 30) support that, for many years the common understanding when selecting reading surfaces has been to avoid colors or backgrounds that are either too light or too dark. In recent years black and other very dark reading surfaces on tables and carrels have appeared. Why this change? This question is worth investigating, because some serious issues are involved. William Pierce, a well-known consultant, cautions against the extremes of light and dark surfaces as well as glare: "The specifies should be cautioned about using top finishes that are too light or too dark. The too-light top is uncomfortable to the eye of the user and difficult to keep clean; the very dark top provides too great a contrast to paper; thus is uncomfortable to the eyes"

According to GCCC (2007), lighting design in library should be considere on 1) Make lighting glarefree with a minimum level of 50 lux at ground level. 2) Eliminate glare, illuminate signage and highlight level changes. 3) Provide uniform illuminance levels internally and comply with the requirements for maintenance illumination in all circulation spaces, including publicly accessible areas.

4) Provide a minimum illumination of 40 lux, uniformity of 0.3 and an average maintained value of 120 lux. Provide a graduated level of illumination at building entries and exits to assist people with vision impairment. Provide a minimum of 50 lux outside the entry or exit.(5) Provide adequate focused lighting for sign language interpretation for people who are deaf/hearing impaired in conference rooms, meeting rooms, auditoria and the like.

3.4. Ambient and psychological aspects of the physical interior environment – affect the human behaviours and perceptions.

It also should also consider on how elements of the physical environment influence human attitudes and user's behaviors. According to Shill & Tonner (2004), points out that some previous studies and surveys on usage and library design have indicated that the design of the physical environment can have significant impact on library usage. Research suggests that the physical setting may also influence the customer's ultimate satisfaction with the service, productivity, and motivation (Bitner 1990;Harrell,Hut, and Anderson 1980). According to Darly and Gilbert 1985;Holahan 1986; Russell and Ward 1982; Stokols and Altman 1987) in Bitner, M. J. (1992), add that human behaviour is influenced by the physical setting and interior environment in which occurs is essentially a truism. Interestingly, the field of environmental psychology has addressed the relationship between human beings and their built environment. Here, it is assumed that dimensions of the organization's physical surroundings influence important customer and employee behaviors. According to (Mehrabian and Russell 1974) cited in

Bitner, M. J. (1992), that environmental psychologists suggest that individuals react to places with two general, and opposite, forms of behavior;approach and avoidance. Approach behaviors include all positive behaviors that might be directed at a particular place, such as desire to stay, explore, work and affiliate. Avoidance behaviors reflects the opposite, in other words, a desire not to stay, explore, work, and affiliate. (Mehrabian and Russell 1974). In other words, the approach of user's behaviors in interior space (including enjoyment, exciting, exploration, attractions, stay longer and returning) are depends on the physical setting in interior environments of libraries.

According to (Bell, 2008) the literature shows that the physical environment influences customer behaviour by:

- Creating strong emotions and mood by means of the sensory qualities of the environment (Grossbart et al. 1990; Baker and Cameron 1996; Sherman, Mather, and Smith 1997; Richins 1997; Le Bel 2005; Berry, Wall, and Carbone 2006);
- Influencing personal interactions between patrons and staff and among patrons (Le Bel 2005);
- Creating a pleasurable total customer experience (Berry, Carbone, and Haeckel 2002); and
- Influencing future patronage intentions (Carbone and Haeckel 1994; Ward and Bitner 1992; Berry, Carbone, and Haeckel 2002; Berry, Wall, and Carbone 2006).

Physical environment includes sensory aspects and has the power to affect mood and emotion. Perception of the environment, in its most strict sense, refers to the process of becoming aware of a space by the acquisition of information through the sensations of sight, hearing, smell, touch, and taste. Cognition is the mental processing of this sensory information. This may involve the activities of thinking about, remembering, or evaluating the information. Spatial behaviour refers to responses and reactions to the environmental information acquired through perception and cognition. The designer creates environmental stimuli to direct these psychological stages as well as the secondary processes of motivation, effect and development. Environmental expectations, another determining element to be considered by the interior designer, are developed over time through experience and interaction with the environment. Sensations, in combination with expectations of the environment, define one's perception of a space. The influence of atmospheric attributes in marketing contexts is based on the premise that the design of an environment through a variety of means such as temperature, sounds, layout, lighting, and colors can stimulate perceptual and emotional responses in consumers and affect their behavior .(Yildirim, Akalin-Baskaya, & Hidayetoglu, 2007)

3.4.1. Ambient influences

According to (Bell, 2008) cited that ambient influences include the non-visual sensory or atmospheric aspects of a service setting, including sounds, smells, and temperature. The literature clearly reveals that these elements have a strong influence in the perception of service quality in hedonic consumption settings (Hirschman and Holbrook 1982; Reimer and Kuehn 2005; Jiang and Wang 2006). Hedonic consumption is based on appeal to the senses and the quality of the sensory input is directly correlated with the perception of service quality. Because these atmospheric influences can arouse intense emotions and create mood, they strongly contribute to the total customer experience (Carbone and Haeckel 1994; Le Bel 2005).

According to Bitner (1990), she points out that, although ambient influences are important in hedonic situations, they are only weakly influential in utilitarian settings. Because utilitarian consumption experiences are based on functionality and sensory influences do not typically affect functionality, the use of sensory stimuli in a utilitarian setting is at best peripheral to an assessment of service quality. Ambient music, for example, can be used to create a pleasant background in a utilitarian service setting, but this element would not be likely to strongly factor into a customer's patronage decision. Ambient influences work in tandem with the visual aspects of the physical setting to create a complete interior environment Ambient stimuli, by their ability to arouse emotions, are a major influence on the psychological aspects of the service setting and can provide strong clues to service quality (Bebko, Scuilli, and Garg 2006; Berry, Wall, and Carbone 2006).

3.4.2. Psychological influences

Psychological influences include emotion, mood, and attitude. Reference to emotions and mood permeate the service quality literature and customer satisfaction and perception of service quality are strongly affected by consumers' psychological reactions (Gardner 1985; Hirschman and Holbrook 1982; Reimer and Kuehn 2005; Jiang and Wang 2006). The interior environment has a powerful ability to generate emotions and those emotions influence customers' perceptions of service quality (Baker and Cameron 1996; Sherman, Mather, and Smith 1997; Richins 1997; Le Bel 2005; Berry, Wall, and Carbone 2006). This is particularly true in hedonic service settings. Although emotions can have an effect on utilitarian encounters, this influence is weak. Consumers' psychological reactions are closely related to the ambient input available at the service site and the physical qualities of the service environment. Psychological influences are also related to the sense of time and to social interactions.

3.4.3. Important role of public library as a social space

According to (Bell, 2008) he cited that likewise, today's library encourages more social interactions within the library and offers a range of group and children space. These changes in library use have forced librarians to rethink their approach to the planning and design of the library building. Many writers emphasize the important role that the library (particularly the public library) plays as a social space, and how space planning must reflect this. Lang (1999) argues that "libraries throughout history have been extraordinarily successful socialspaces" (11), and others agree that the library can help to satisfy a "yearning for informal socialization" (Lushington, 2002:91) and is important as "a public space and meeting place" (Dewe, 2006:21).

The increasing importance of this role, particularly for public libraries, usually results in both openplan layouts in the main part of the library that maximise contact between different elements of the community and also separate facilities like cafes and meeting rooms for community groups which can help preserve a quiet atmosphere in other parts of the building (Dewe, 2006; Greenhalgh et al., 1995; Lushington, 2002; Sannwald, 2007). In his account of the development of the Lanchester Library Noon stresses that they were creating "a kind of shopping mall where the main attraction was... the Library!" (Noon, 2004:94).

Furthermore, nowadays public libraries becoming learning space and at the same time becoming a place where users come to acquire knowledge, communicate, socialize and take a rest. The new ideas of adding a place for refreshment such as a café or small coffee bar in the public libraries are becoming an issue. Some young people would like to visit the library as they are visiting coffee-shops – have relaxed atmosphere where they can chat, study or have a rest (Pomerantz & Marchionini, 2007).However, as Hurt (2000) admits, this means going against beliefs that are deeply held by many librarians, though she was glad that at George Mason University they "broke some rules that should no longer be automatically accepted" (103), including allowing food into the library. According to Dewe (2006), these and other taboos are being increasingly challenged in 11 public libraries too as "rules about eating and drinking have been relaxed and noise disregarded" (23).

3.5. Understanding on the effects of the physical environment and its relation in producing a better interior ambiances design

Understanding on the effects of the physical environment and its relation in producing a better interior ambiances design can be found in Figure:1. Physical interior environment is one of the most important aspects in enhancing better design in libraries. The researcher found that there are four physical interior environments should be emphasized in providing a better design in libraries such as that the selection of

- Lighting - day lighting (natural lighting) or artificial lighting. The selection of lighting is based on a standard comfortable lux and luminance.
- Furniture - selection of furniture that is appropriate according to the ergonomics of children, adults or a senior citizen. It also should be according to the needs of the group or individual user, a selection of furniture such as types of seating, reading tables, bookshelves and it also must be ergonomic and appropriate to each user.
- Space planning – the interior layout and space requirements,
- Materials and finishes - choice of colours and finishes are appropriate to the furniture, wall, floor and ceiling.

The researcher also found that there are some aspects that affect emotions, mood and user's experiences. It also influenced human behaviors and perceptions through aspects of ambient, aesthetics and ergonomic factors. Those aspects affect users to come back to the library, to stay longer and find that the library is a fun and exciting place to explore and visits.

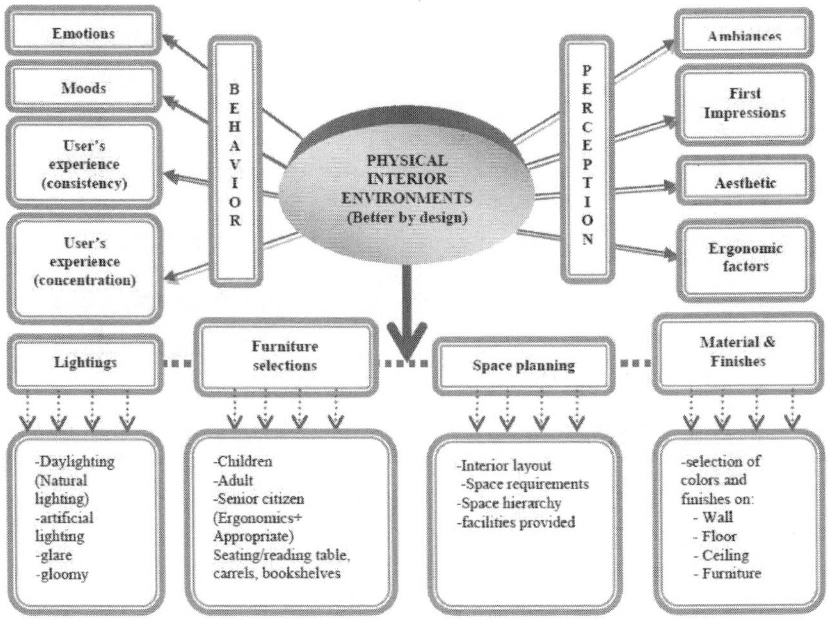

Figure 1. Understanding on the effects of the physical environment and its relation in producing a better interior ambiances design.

There are also some comparisons of old and new approaches of the physical library interior aspects of past and present, or the digital age. Summary of the characteristics of the old and new of the library interior are included in Table 1. The comparison is carried out by using observation of the libraries and doing source research (Anandasivam, K. & Cheong, C. F. 2008; Childs, P. 2006;

Hulse, P. 2007; Pomerantz, J. & Marchionini, G. 2007; Tseng, S. 2008) cited by Juhnevica & Udre, (2010). It can be described with the research finding studied by several criteria to be considered, include the concept of the library, the interior space, lighting, colors, finishes and furniture selection.

Table 1. Comparison of old and new approaches of the physical library interior aspects. Source: (Juhnevica & Udre, 2010)

Criteria	Old libraries	Current/digital age libraries
Concept of the library	Silent place for studying and working	Active place for retrieving information and socializing with other users
Space in the library	Much closed space, furniture close to each other	A lot of open space, space to move around between furniture
Rooms in the library	Big silent reading rooms, separated computer rooms	Big active reading rooms to use personal laptops with free Wi-Fi Internet access, and socialize with other users; separated silent reading rooms; additional computers in the library; special rooms for group-works
Lighting	Dark reading rooms with little lamps and extra lamps on reading desks	Bright reading rooms with a lot of light and extra lamps on reading desks. Design features where appropriate are the use of atria and skylights and using glass windows for maximizing the use of natural lightings.
Colours	Dark colours (brown, gray, beige, gold, deep red, deep green)	Bright colours (orange, blue, yellow, green, white, silver)
Furniture material	Heavy wood, chipboard	Plastic, textiles
Desks	Heavy, unmovable, made of wood	Light, easy to move, made from different materials
Chairs	Heavy and hard wooden chairs, may be with leather seats	Easy portable and comfortable chairs with soft padding
Shelving	Big and heavy shelves, very tall ones with additional ladders	Easy-to-reach shelves, movable shelves, automatic shelves

4. CONCLUSION

It is important to understanding of the role of the physical interior environment on users' perceptions of creating a better design. In researcher opinions and understanding, to achieve a better design of public libraries, aspects of physical and environmental criteria should be emphasized from the beginning of the planning space.

By understanding all of the influences on the perception and expectation of users on creating a better design, decision makers can ensure that their approaches to those ambient and physical elements within their expertise appropriately interact with and enhance the other influences on the total user's experience and behaviors. It is more than just the functionality and aesthetics of the ambient and physical design, but also temporal affects, social interaction and psychological consequences. Therefore, to create and to achieve a better design for future public libraries, the various aspects of physical interior environments and interior ambience should be consider such space planning, selection of furniture, lightings, material and finishes. According to (Bell, 2008), the interior environment comprises two of these influences – physical and ambient – and strongly affects the remaining functional, temporal, emotional, indicative, and social influences. Each of these influences affects customer response both directly and indirectly through its affects on the other influences. The physical setting is one of the most prominent and pervasive of these influences. In many ways, this statement sums up the service quality literature as it has evolved to its present state. Although physical environments comprises many factors, including the ambient, interior environment, that interact and influence each other in numerous ways, ultimately, the focus must be on the users – his or her expectations, desires, needs, and experience.

REFERENCES

1. (PNM), P. N. (2001). Garis Panduan Perpustakaan Khusus. Kuala Lumpur, Malaysia: Ampang Press Sdn. Bhd

2. Anandasivam, K., & Cheong, F. C. (2008). Designing a Creative Learning Environment: NTU's New Art, Design & Media Library. The Electronic Library Oxford , vol. 26, page 650.

3. Bell, C. R. (2008). The Role of the Interior Environment in the Perception of Service Quality:A Business Perspective. Marymount University, Faculty of the School of Arts and Sciences, Arlington, VA.

4. Bitner, M. J. (1990). Evaluating service encounters: the effects of physical surroundings. and employee responses. Journal of Marketing , 54 (2): 69.

5. Bitner, M. J. (1992). Servicescapes: the impact of physical surroundings on customers andemployees. . Journal of Marketing , 56 (2): 57.

6. Clemons, S. A. (2009). University Libraries as Third Place. IFLA World Library & Information Congress:75th IFLA General Conference & Assembly. Torino, Milan, Italy.

7. Dewe, M. (2006). Planning public library buildings : concepts and issues for the librarian. Hants, England: Ashgate Publishing Limited.

8. Economic Planning Unit & Costs, E. P. (2008 Edition). Guidelines and Rules for the Planning and Building Standards Committee . In SECTION G-Library. Putrajaya, Malaysia.

9. Eigenbrodt, O. (2009). Libraries and Learning Centres; Current Approach. IFLA: Section on Library Buildings and Equipment .

10. Eigenbrodt, O. (August 2009). Physical Space in Theory and Beyond: Building Libraries for the Knowledge Socities. World Library and Information Congress:75th IFLA General Conference and Assembly. Milan, Italy.

11. Environment), C. (. (2003, August). Better Public Libraries . Retrieved from www.cabe.org.uk/files/better-public-libraries.pdf .

12. Francine, M. (2009). Theme: Libraries as Place and Space ,In the Words of the Users: The role of the urban public library as place. World Library and Information Congress: 75th IFLA General Conference Assembly. Torino,Milan, Italy.

13. Gold Coast City Council Branch Libraries (GCCC). (2007). Generic Design Brief for Gold Coast City Council Branch Libraries (GCCC). Gold Coast, Australia.

14. IFLA. (2009). Libraries and Learning Centre; Current Approaches. News 2009 No. 1 Section on Library Buildings & Equipment .

15. IFLA. (10-14 August 2008). Library as space and place: theoritical approach. World Library and Information Congress:74th IFLA General Conference and Council. Quebec, Canada.

16. Jiang, Y. a. (2006). The impact of affect on service quality and satisfaction: the moderation of service contexts. Journal of Services Marketing , 20 (4): 211.

17. Juhnevica, E., & Udre, D. (2010). "Libraries Designed for Users" Nowadays Concept of Library Architecture and Interior. BOBCATSSS 2010 @ Parma, Italy .

18. Khan, A. (2009). Better by Design: An Introduction to Planning and Designing a New Library Building. London: Facet Publishing.

19. Kingsbury, A. (2007). To What Extent Do New Library Buildings Meet The Needs of Users and Staff? THE UNIVERSITY OF SHEFFIELD, Master of Arts in Librarianship.

20. Lamar, V. (1987). Toward the Environment Design of Library Buildings. Library Trends , 361-376.

21. Latimer, K. (2007). USERS AND PUBLIC SPACE: WHAT TO CONSIDER WHEN PLANNING LIBRARY SPACE. Berlin, New York (Walter de Gruyter – K. G. Saur): IFLA Library Building Guidelines: Developments & Reflections.

22. Matthews, G. W. (19-21 August 2009). Libraries as space & place:The effective use of space in university and public libraries; current issue. IFLA World Library & Information Congress:75th IFLA General Conference & Assembly. Torino, Milan, Italy.

23. McCabe, G. B. (2003). Planning the Modern Public Library Building. London: The Libarries Unlimited Library Management Collection, Greenwood Publishing Group.

24. McCabe, G. B., & Kennedy, J. R. (2003). Planning the Modern Public Library Building. London: The Libraries Unlimited Library Management Collection.

25. McCabe, G. (2000). Planning for A New Generation of Public Library Buildings. London: Greenwood Press.

26. Olaf, E. (2009). Physical Space in Theory and Beyond: Building Libraries for the Knowledge Societies. IFLA World Library & Information Congress:75th IFLA General Conference & Assembly. Torino, Milan, Italy.

27. Olaf, E. (2009). Physical Space in Theory and Beyond: Building Libraries in the Knowledge Societies. IFLA World Library & Information Congress:75th IFLA General Conference & Assembly. Torino, Milan, Italy.

28. Rooney-Browne, C. (2009). Libraries as space and place (Public Libraries as impartial spaces in the 21th Century). IFLA World Library & Information Congress:75th IFLA General Conference & Assembly. Torino, Milan, Italy.

29. Sannwald, W. W. (2001). Checklist of library building design consideration (4th Edition ed.). Chicago: American Library Association.

30. Shill, H. B., & Tonner, S. (2003). Creating a Better Place: Physical Improvements In Academic Libraries, 1995-2002. College & Research Libraries , 64(6), 431-466.

31. Shill, H. B., & Tonner, S. (2004). Does the Building Still Matter?Usage patterns in new, expanded and renovated libraries,1995- 2002. College & Research Libraries , 65(2), 123-150.

32. Yildirim, K., Akalin-Baskaya, A., & Hidayetoglu, M. (2007). Effects of indoor color on mood and cognitive performance. Science Direct , 42 (9), 3233-3240.

9

CHAPTER

Entrepreneurship Education in Malaysia's Public Institutions of Higher Learning—A Review of the Current Practices

Mohd Nor Hakimin Bin Yusoff[1] , Fakhrul Anwar Zainol[2] & Mohamed Dahlan Bin Ibrahim[1]

[1] University Malaysia Kelantan, Pengkalan Chepa, Kota Bharu, Kelantan, Malaysia

[2] University Sultan Zainal Abidin, Kuala Terengganu, Terengganu, Malaysia

ABSTRACT

The need for a practical and applicable model for entrepreneurial learning is becoming critical. In this study, we aimed to collect data related to entrepreneurship education practices by all institutes of higher learning (IHLs) in Malaysia as well as challenges faced, facilities, and supports offered by the universities. Given the important role of IHLs to nurture and to create future entrepreneurs, the government via public IHLs aspires to produce "job creators" rather than "job seekers." The findings of the study are significant as they provide practical insight into the entrepreneurship education practiced by the local public universities. In addition, the study also contributes to the present literature on the entrepreneurship education in Malaysia, particularly in presenting the practical issues and challenges faced by the universities in conducting entrepreneurship education programs.

Keywords: entrepreneurship education, enterprise education, public university

1. INTRODUCTION

The need for a practical and applicable model for entrepreneurial learning is becoming critical. In this study, we aimed to collect data related to entrepreneurship education practices by all institutes of higher learning (IHLs) in Malaysia as well as challenges faced, facilities, and supports offered by the universities. Altogether Malaysia has 20-government funded universities across the country. The government is divided into 14 regions, which are called states. Each of these 14 states has at least one public university; however, few states have more than one public university. Permodalan Nasional Berhad (National Capital Limited) funded this study.

Generally, people believe that entrepreneurial skills can be learned by inculcating entrepreneurship knowledge to students. Supposedly, students with entrepreneurship knowledge are more inclined towards venturing into business than are those who have not been exposed to this knowledge. For that reason, entrepreneurship education has turned out to be the central focus of the education ministry, and educators are encouraged to promote an entrepreneurial learning pedagogy that combines classroom theoretical lectures and practical learning outside the classroom to obtain a good impact and an enriched learning experience. Furthermore, students are exposed to the self-directed learning together with problem-based learning, which is proven to have a significant impact on students' entrepreneurial intentions.

Nevertheless, some educators still oppose the idea of students venturing into the business while they are still pursuing their studies (Fayolle & Gailly, 2008). They have reasoned that business activities may interfere with students' learning time and affect students' focus on their studies. They have also argued that spending more time on business activities may contribute to students' psychological and physiological problems due to fatigue. Despite these educators' criticisms, most universities believe in the benefits of entrepreneurship education and have aggressively promoted the subjects

Scholars have different views in defining entrepreneurship education. For universities in Finland, this education is called enterprise ownership and intrapreneurship, and it is more inclined to the traditional way of defining entrepreneurship. In Great Britain and Ireland, entrepreneurship education is called enterprise education. The curriculum content is more focused on individual capacity building, i.e. entrepreneurial attributes. Therefore, the establishment of the business entity is not the ultimate outcome of the curriculum (Nurmi & Paasio, 2007). The diversity of concepts and opinions provides the opportunity for researchers to conduct empirical studies to narrow the gap in the existing relevant literature. The findings from this particular study will also provide a basis to develop a new entrepreneurship education model that considers the deficiencies in the current model. The proposed model would become a blueprint and act as guidelines for all IHLs to implement effective entrepreneurship education programs and activities.

Our main objective for this study was to collect data about the entrepreneurship programs conducted and support provided by Malaysia's public IHLs. The study is also aimed to identify challenges faced by the universities while promoting their entrepreneurship education programs. Given the important role of IHLs to nurture and to create future entrepreneurs, the government via public IHLs aspires to produce "job creators" rather than "job seekers." The findings of the study are significant as they provide practical insight into the entrepreneurship education practiced by the local public universities. In addition, the study also contributes to the present literature on the entrepreneurship education in Malaysia, particularly in presenting the practical issues and challenges faced by the universities in conducting entrepreneurship education programs.

2. METHODOLOGY

For this study, we used a qualitative approach where the data were gathered via structured, personal interviews and open-ended questions. We selected this stratified sampling approach and the respondents from each of the 20 public universities. The respondents were personally responsible in the administration, supervision, and coordination of entrepreneurship programs at their respective universities. The data gathered were analysed using a descriptive comparison technique.

The questionnaires were divided into six parts. Part A was about the university's entrepreneurship development model. The respondents were required to answer on the availability of such model and whether it is documented as a policy paper. Most of the questions in Part A consisted of dichotomous questions that required respondents to give a yes or no answer. Part B consisted of four questions. The first question was about whether each respondent's school had a compulsory entrepreneurship course. Secondly, respondents reported the credit hours allocated for the course (if any). The third question required respondents to complete the information on the programs undertaken by the university such as the "Students Entrepreneurship Training

Series" or "Modules and Students Basic Entrepreneurship Course." We also requested that the respondents furnish information on the entrepreneurship co-curriculum activities at their universities, including exhibitions, ad-hoc sales stalls, community services, tea-talk, and promotion, among others. The fourth question addressed the issue of the lecturers' qualifications. We asked respondents to identify an area of expertise for a skilled lecturer. Part C included questions about the entrepreneurship education's targeted group and the university's policies regarding student business activities.

In Part D, the questions were about the entrepreneurship support programs available in the respondents' respective IHLs. We asked respondents to identify the availability of support, the type of support, and the programs held to promote entrepreneurship education programs. The questions in Part E were about the effectiveness of the entrepreneurship curriculum practiced by the IHLs. Respondents responded to questions about the effectiveness of the curriculum and the number of graduate entrepreneurs that university had produced. In addition, we asked respondents to list the problems and challenges faced by students, and then to propose solutions to overcome the problems. Finally, Part F included questions to gather the information pertaining to the respondents' demographic profiles, such as age, education levels, areas of expertise, locations, yeasr of services, and employers' addresses.

3. LITERATURE REVIEW

Entrepreneurship plays an important role in promoting economic activities in the country by creating job opportunities, and it contributes significantly to the country's gross domestic product (National SME Development Council, 2007). Considering the significant contribution to the economic development and prosperity of the country, the government has intensified its effort in promoting entrepreneurship activities among its citizens. Besides acting as a catalyst to wealth creation, entrepreneurial activities also bring benefits in terms of less dependence on the government in providing job opportunities. Thus, nurturing the mental attitude towards entrepreneurial activities among students can be done through the entrepreneurship education process (Gorman, Hanlon, & King, 1997).

3.1 Nurturing Entrepreneurship Culture via Formal Education

Previous studies have shown that entrepreneurship is a discipline similar to any other discipline and therefore it can be learned. A person is neither born nor fated to be an entrepreneur. Researchers have expressed the belief that entrepreneurs can be created by way of effective education, training, and the nurturing process. Drucker (1994) posited that: Most of what you hear about entrepreneurship is all wrong. It's not magic, it's not mysterious; and it has nothing to do with genes. It's a discipline and, like any other discipline, it can be learned.

Based on the above argument, the entrepreneurship programs offered by local public universities either in the form of a compulsory core subject or embedded in the curriculum is aimed to produce as many graduate entrepreneurs as possible. Nevertheless, the impact has not been satisfactory if the program has failed to attract graduates to venture into entrepreneurship activities immediately after graduation. This phenomenon attracts sceptics who criticize public universities for failing to produce graduate entrepreneurs, while other institutions, such as technical and other vocational training centres, have succeeded. A study conducted by Cheng, Chan, and Mahmood (2009), revealed that the effectiveness of entrepreneurship education in Malaysia is correlated with the content of the curriculum. The study revealed that the content of the present curriculum has not met the requirements and that there is a clear mismatch with the students' expectations.

3.2 The Essentials of Entrepreneurship Education

The main objective of entrepreneurship education in public IHLs is to produce entrepreneurial graduates as well as graduate entrepreneurs. Entrepreneurship education has been promoted based on the belief that entrepreneurship can be nurtured and learnt. In addition, continuous exposure to entrepreneurship activities can enhance students' inclination to venture into these activities. In addition to a students' acquiring the entrepreneurship knowledge and experience, their family backgrounds, personal experiences, and external environment were also found to influence their intentions towards involvement in entrepreneurship activities. People with a family background in business tend to have higher inclinations towards entrepreneurial activities compared to those who do not (Dahlan, Hakimin, & Azwin, 2010). Early and continuous exposure to entrepreneurship activities have inculcated the tacit knowledge that was found to positively influence a person's intention and behaviour to venture into business in particular. On the other hand, students who have been exposed to entrepreneurship education were found to positively correlate to an inclination towards venturing into business. They possessed entrepreneurial characteristics and act entrepreneurially compared to other student without similar exposure (Cheng et al., 2009). Thus, entrepreneurship education becomes vital to nurturing and encouraging students to choose entrepreneurship as a career option. Therefore, the number of graduate entrepreneurs will likely increase if more students are exposed to entrepreneurship education programs and activities in the university. Curriculum, however, should be designed to maximize students' learning experience via the entrepreneurship education and enterprise education models. Conversely, students who have previous working experience could perform entrepreneurially as they already possess many entrepreneurial characteristics. Under the current employment scenario, prospective employers prefer students who can act entrepreneurially as they are able to deal with the dynamics of the business environment (Tan & Ng, 2006).

3.3 The Delivery Method

Most of the existing entrepreneurship education curriculum emphasises the theoretical knowledge and gives less attention to the practical application of the subject. The programs accentuate the established knowledge; however, they are frail on skill development and tacit experiences, which are critical features for nascent entrepreneurs (Dhliwayo, 2008). Presently, in the conventional way of teaching entrepreneurship, students are exposed to the philosophy and theory with minimum exposure to the practical aspects. The applications to imagination, creativity, and innovation have only been theoretically explained, and as a result, students have been unable to emulate and grasp the experience of being an entrepreneur. This particular shortcoming needs to be revamped as its failure entails serious implications on government expenditures because it involves a huge amount of money and wastes a lot of time. In addition, an urgent review of the conventional approach is needed as it does not emphasise the lecturer's abilities within the entrepreneurship subject. Such a review is more critical for cases when the lecturer has insufficient experience and knowledge about entrepreneurship. To overcome such a problem, the entrepreneurship educators' competency should be intensified via continuous learning and competency-building through business sabbaticals in order for them to gain more experience that would enhance their teaching effectiveness. Experiential learning is a pedagogical approach that enhances the process of knowledge acquisition via an experience-transformation process. This approach postulates that effective learning can only take place when experience is acquired (Dhliwayo, 2008).

Due to the weaknesses that have prevailed in the current teaching techniques, many researchers have introduced a number of teaching approaches in entrepreneurship education. New learning models have been promoted to enhance the effectiveness of the process of nurturing and motivating students towards entrepreneurial activities. A South African University has come out with the "Entrepreneurial Performance Education Model" (EPEM) and the "Entrepreneurial Education Model" (EEM) (Van Vuuren & Nieman, 1999). EPEM emphasizes motivation, entrepreneurial skills, and business skills that influence entrepreneurial performance. EEM takes into account the fact that the facilitator and the teaching approach are essential elements in the education program (Pretorius, Nieman, & van Vuuren, 2005).

Cheng et al.'s (2009) study showed that university students have indicated no significant relationship between entrepreneurship knowledge and intent to venture into business. The study revealed that the current entrepreneurship education has failed to enhance students' inclination to choose entrepreneurship as a career option. Respondents gave such feedback as the following: inexperienced lecturers, ineffective methodologies that are examination oriented and too theoretical, and insufficient practical exposure. Cheng et al. (2009) also found that 84% of the programs were conducted via lectures, 11.5% by way of case study, and 6.3% lectures were delivered by guest speakers. The study showed that most of the entrepreneurship learning process was done through classroom teaching, which is theoretical in nature. The findings suggested that entrepreneurship philosophy was not well delivered because the majority of the respondents perceived entrepreneurs as being born and therefore an average person could not be trained. Such a myth may drive students to assume that nobody can turn him or herself into a successful entrepreneur even when acquiring entrepreneurship education (Jones & English, 2004), or that no one can become an entrepreneur unless he or she were born to be one.

4. FINDINGS

We have divided the findings of the current study into three subsections: (1) findings about the support according to each university's authority and type of support, (2) related activities, and (3) the respondents' challenges and suggestions to resolve such problems.

Malaysia has 20 public universities located across the country, and each state has at least one public university. The study found that only 1 out of the 20 universities has not introduced any entrepreneurship program–Universiti Pertahanan Malaysia (Defence University of Malaysia). The reason for this lack is that this university's educational objective is to produce high ranking army personnel, thus they have nothing to do with entrepreneurship. The survey results showed that 19 public universities have indicated that they carry out entrepreneurship programs as a core subject. Details of the finding are in the following subsections.

4.1 Support

Table 1 below summarizes the various types of support that may enhance the success of entrepreneurship programs executed in Malaysia's public universities.

4.1.1 The Management Support

85% of the respondents agreed that the entrepreneurship programs received strong support from their universities' top management. The balance of 15% claimed that their entrepreneurship programs are not well supported by the top management.

4.1.2 Infrastructure Support

80% of the respondents agreed that the infrastructure for entrepreneurship activities were available in the universities while 25% of the respondents claimed that their universities had insufficient infrastructure available to support the entrepreneurship programs.

4.1.3 Promotion, Funding, and Advisory

25% of the respondents claimed they did not conduct a comprehensive promotion program to promote the entrepreneurial activities and programs among students. Six universities were found to have no provision for funding students' business activities while seven universities did not have external experts to advise on the conducted programs.

Table 1. List of support offered

Univ Malaysia Kelantan	√	√	X	X	√
Univ Teknoloji Malaysia	√	√	X	X	√
Univ Teknologi Hussin Onn	√	√	√	X	X
Univ Sains Malaysia	√	√	√	√	√
Univ Malaya	√	√	√	√	√
Univ Kebangsaan Malaysia	√	√	√	√	√
Univ Putra Malaysia	√	√	√	√	√
Univ Teknologi MARA	√	√	√	√	√
Univ Malaysia Perlis	√	√	√	√	√
Univ Utara Malaysia	√	√	√	√	√
Univ Malaysia Pahang	X	X	√	√	X
Univ Sultan Zainal Abidin	√	X	X	√	X
Univ Malaysia Terengganu	√	X	√	√	X
Univ Perguruan Sultan Idris	√	√	√	X	√
Univ Pertahanan Malaysia	X	X	X	X	X
Univ Sains Islam Malaysia	√	√	√	√	√
Univ Teknikal Melaka	X	X	X	X	X
Univ Malaysia Sabah	√	√	√	√	√
Univ Malaysia Sarawak	√	√	√	√	√
Univ Islam Antarabangsa	√	√	√	√	√

4.2 Activities

4.2.1 Entrepreneurship Development Model

Most of the universities have adopted an education model for entrepreneurship programs as outlined by the Ministry of Education (MOE). However, three universities have applied their own models. The IHLs are Universiti Malaysia Kelantan, Univversit Utara Malaysia, and Universiti Teknologi MARA.

4.2.2 Entrepreneurship Courses

All respondents agreed that, in order to adhere to policy stipulated by MOE, their university have made an entrepreneurship course compulsory for all students. However, each IHL has slight differences in terms of the number of credit hours given to students, including between two and three credits.

4.2.3 Students Entrepreneurship Training, Entrepreneurship Students Basic Course, and Graduate Entrepreneur Development Programs

All of the programs are implemented at each of the public IHLs except the National Defence University of Malaysia. UMK, however, has used different terminology and approaches in implementing these programs.

4.2.4 Entrepreneurial Activities

The entrepreneurial activities that include seminars, workshops, community service, entrepreneurship clubs (or a combination of these) have been implemented in all IHLs as extra-curricular activities. Additionally, some IHLs have been implementing their own programmes, as shown in the Table 2

Table 2. Internally developed programs

No	IHL	Program/Activity
1	UMK	PELS, CLS, SIEP, SEED, SEC.
2	USM	SIFE, SEDIA, Technopreneurs
3	UiTM	SIFE, Technology entrepreneur
4	UniMAP	*Business Game*
5	UTHM	Ko Siswa
6	UTM/UUM	*Student's Mall*
7	UMS	Business and Economy School Week, Workshop, Entrepreneurship Program

Note: PELS–Prime Entrepeneur Lecture Series, CLS–Chairman Lecture Series, SIEP–Students in Enterprise Programs, SEED –Social Enterprise and Economic Development, SEC–Student Enterprise Center, SIFE–Student in Free Enterprise

4.2.5 Entrepreneurship Educators

In terms of delivery, respondents reported that half of the IHLs employ a combination of people from academic and professional/industrial backgrounds. However, some lecturers with no entrepreneurship education qualification and background have been appointed as trainers.

4.2.6 Target Groups

Participants reported that all IHLs' students are the targeted groups. Almost all IPTAs set target entrepreneurship education to all students, irrespective of their discipline of study. In addition, a number of IHLs have extended the scope of entrepreneurship education to include their alumni, the rural community, and post-graduate students.

4.2.7 University Policy

Fourteen public IHLs have set up policy governing the students' entrepreneurial activities, academic achievements, and students' business loans.

Out of 21 IHLs, 19 have received support in various forms for their entrepreneurship activities. The supports include help with management, infrastructure, promotion, financing, and advisory services. Furthermore, eight IHLs claimed that they are practicing scheduled monitoring on the effectiveness of the entrepreneurship education in their respective IHLs.

The findings showed that only five IHLs (UKM, UUM, UiTM, USIM and UMS) have undertaken a study on their entrepreneurship education effectiveness. Almost 50% of the IHLs have maintained and kept track of the number of their graduates who have successfully ventured into business.

The above findings indicate that entrepreneurship education has been one of the core programs in Malaysian public IHLs. However, they also indicate various approaches in the implementation of the programs. Such variance is due to the degree of emphasis the respective IHLs have given to entrepreneurship education. Therefore, Malaysia IHLs have need for a study to compare and to assess the various models currently being applied in the IHLs so that a more comprehensive entrepreneurship education model can be developed and implemented in all public IHLs.

Table 3. List of challenges

	Most frequently mentioned
Challenges	1. Students soft skill 2. Financing 3. Commitment 4. Coordination among support agencies 5. Bureaucracy 6. Manager support 7. Staff and student mentality 8. Pedagogy

4.3 Challenges Faced by IHLs

According to the feedback, the majority of the IHLs recognize the current challenges faced in executing entrepreneurship education into the present curriculum. The challenges include lacks in soft skills, in financing, and in commitment from all parties. At the top of the list was a lack of soft skills among the students. Students

Table 4. Challenges and suggestion of solutions

Institute of Higher Learning	Challenges	Recommendations
Univ Malaysia Kelantan	X	X
Univ Teknoloji Malaysia	Lack of interest among students, capital funding, experience, skill, financial management, coordination, space	Space, grant, Mentor-mentee, Skill development, subsidy
Univ Teknologi Hussin Onn	Financing, networking, lack of knowledge, mentality	Standardised term of financing, financial knowledge, cooperation with SMIDEC and SME Bank.
Univ Sains Malaysia	Credit bureaucracy	Bridging arrangement
Univ Malaya	Staff and student lack of commitment	Awareness programs
Univ Kebangsaan Malaysia	Leadership, communication	More time, expert.
Univ Putra Malaysia	Mentality	Motivation
Univ Teknologi MARA	Funding	Dedicated fund, continuous business opportunity
Univ Malaysia Perlis	X	X
Univ Utara Malaysia	Funding, lack of information, networking, Mentality, coordination among agencies	Integrated program, agency proactive, networking creation, support
Univ Malaysia Pahang	Coordination, no special unit, continuity	Continous support
Univ Sultan Zainal Abidin	No expertise. Bureaucracy, no support from higher authority	TOT, establishment of special unit
Univ Malaysia Terengganu	No special unit	Coordination
Univ Perguruan Sultan Idris	Pedagogy	More time, entrepreneurship related subject
Univ Pertahanan Malaysia	X	X
Univ Sains Islam Malaysia	Culture nurturing	X
Univ Teknikal Melaka	X	X
Univ Malaysia Sabah	Lack of confidence, funding, experience, marketing strategy	Curriculum revamp, continuous guidance from entrepreneurs
Univ Malaysia Sarawak	Funding, lack of experience	Friendly procedure, enhancement of training and practice
Univ Islam Antarabangsa	X	X

Note: X–None.

following the entrepreneurship programs lacked communication skills, leadership skills, knowledge of marketing strategies, general business knowledge, and abilities in networking. The second top challenge cited was the difficulty of students to obtain loans to start their business on campus. The third top challenge was lack of commitment among IHLs' staff and students. This feedback came from five IHL participants who found that staff and students were not committed to the entrepreneurship programs. The list of challenges here is a based on the rankings of the most frequently mentioned challenges, as noted in Table 3.

Other challenges include the issue of coordination among support agencies, bureaucracy, and lack of management support, unfavourable mentality among staff and students towards entrepreneurship activities, and ineffectual pedagogy.

5. DISCUSSIONS

The majority of Malaysia's public IHLs provide support for entrepreneurship education. The most common and popular form of support is the provision of business premises to allow students to run entrepreneurship activities. However, IHL managements do not cover some of the most critical in-business field activities for unknown reasons. The findings of this study show that a substantial number of IHLs face difficulties securing seed funding for student enterprises. In spite of this difficulty, some IHLs do offer seed capital in the form of soft loans, but the size of the funds is small. The amount allocated for seed capital by the university is small and managed by way of a revolving fund. Because they are revolving funds, their availiability depends on the repayment collected from the current borrower. The funds have constantly experienced a shortage since repayment is slow and erratic, and, as a result, the newly set up enterprise fails to secure the loan. In the worst case scenario, some IHLs experience high non-performing loans and have to stop the loan facilities. This is a common scenario happening when discussing entrepreneurship education in most IHLs. The findings of the study implicitly indicate that some IHLs are not serious in running and providing the necessary entrepreneurship support programs. Entrepreneurship is not all about selling products; the scope is wider to cover all business skills, including the individuals' hard and soft skills. Furthermore, many public IHLs do not have clear guidelines or policies related to their entrepreneurship education and activities. In the absence of student entrepreneurship policies and effective advisory services, the entrepreneurship programs will not be effective in achieving the intended goals. It is imperative for public IHLs to have clear policies as students may engage in business activities that have legal implications on them and on the university. Policies need to provide guidelines on actions that should be taken in the event that students encounter legal problems, so that such encounters will not negatively affect their studies.

Another significant finding from this study is about the trainers' skills. Some IHLs have simply been appointing trainers without considering their educational backgrounds and experiences. The results of this study indicate that some of the trainers have no qualifications to conduct entrepreneurship programs due to irrelevant education background and experience. This is a fundamental and crucial issue to ensure the success of these university's entrepreneurship programs. This phenomenon indicates that some IHLs' management do not see entrepreneurship education as a significant focal point. The management should be aware that entrepreneurship education is not only about learning the theories of business, but also it demands extensive exposure to personal practical experience, an effective pedagogical method, and an experienced and competent trainer.

The study's findings also indicate that the public IHLs have a mixed understanding of the objectives of entrepreneurship education. The majority of IHLs believe that that entrepreneurship education is to inculcate entrepreneurial characteristics to add value to graduates and to enhance the graduates' chance to attain employment. On the other hand, some

IHLs' objective of entrepreneurship education is to produce entrepreneurs among their graduates. Different objectives require different pedagogy for the programs to be successful. Mixed understandings of the objective affects the uniformity in the implementation of the programs. The time has come for the highest authority, i.e. MOE, to introduce a new entrepreneurship education model that is more effective and that integrates both objectives.

Regarding the examination of the pedagogy applied by the public IHLs, the findings of this study show that the entrepreneurship education delivery system is less effective. Most of the teaching approaches are generic and fail to inculcate entrepreneurship's true value. The findings also indicate that lectures are the most common and popular methodology in the delivery of entrepreneurship education, over methods such as business simulation, role play, and case studies. Teaching methods that are teacher-centred may be appropriate in the context of explaining theoretical concepts; however, such a method is not able to provide actual experience and consequently the knowledge fails to be put into practice and students' learning experience is not maximized.

In terms of the impact of the programs, the findings indicate that there has not been a significant impact on graduates' tendency to choose entrepreneurship as a career. IHLs have invested a considerable sum of money in providing the resources needed for the implementation of entrepreneurship education programs and activities; however, the percentage of graduates who want to take up entrepreneurship as a career option is very low and far from achieving the target. The failure to promote entrepreneurship as a career among graduates seems to be caused by the lack of an effective entrepreneurship education model with realistic goals. Currently, most of the entrepreneurship education programs do not have specific learning outcomes and have been implemented in silos. In other words, the programs have failed to support the two prong objectives of entrepreneurship education in Malaysian public IHLs.

To summarize, the methodology used in entrepreneurship education should focus on the entrepreneur's attitudes and attribute development, which were earlier identified, such as internal locus of control, optimism, commitment and confidence, problem solving skills, and risk-taking propensity, among others. Current methods that adopt traditional teaching, such as through a theoretical framework and being examination-oriented, are less practical in efforts to attract students to venture into business after graduation. The problem is compounded when the entrepreneurial class is conducted by instructors or lecturers who are not qualified and never have been exposed to entrepreneurship training. Such a situation is awkward for someone to have to talk about entrepreneurship without having a formal education background and never having taken training or courses to enhance his or her entrepreneurial knowledge. Thus, to ensure the effectiveness of the entrepreneurship education programs, a new method of entrepreneurial teaching and learning and revision of the current curriculum should be undertaken in order to boost the understanding of the entrepreneurial process and the way to nurture an entrepreneurial attitude. The entrepreneurship curriculum needs to include courses that are geared towards the development of students' business acumens and entrepreneurial motivation. Business acumens cover financial literacy and general management, human resources management, communications, legal, and marketing skills. Entrepreneurial skills refers to the ability to assess risk and creativity and the ability to identify business opportunities. Based on the Babson College's practices, experience-based learning can improve skills and motivate students to venture into entrepreneurship. Past studies have shown that learning based on experience is more effective than the traditional way of learning. The time has come for a comprehensive model of entrepreneurship education to be formulated and implemented in order to enhance the effectiveness of entrepreneurial teaching and learning. The success of the entrepreneurship education would be the beginning of a new era of entrepreneurship as a career of choice.

6. RECOMMENDATION

Based on the above findings and discussion, this study suggests the stakeholder to introduce a new entrepreneurship education model that integrates entrepreneurship education and enterprise education. This study proposed the new model which is called Entrepreneurship Education and Enterprise Education Model or in brief the 4e Model. The graphical model is shown in Figure 1.

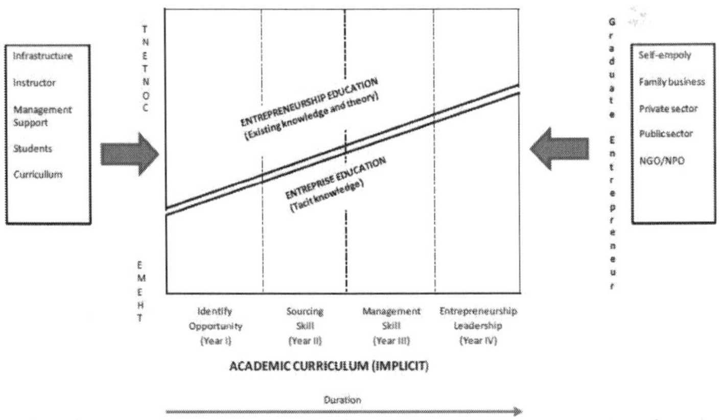

Figure 1. Education and enterprise education model (The 4e Model)

The proposed model adopts the input-transformation-output process of entrepreneurial learning. The transformation process where the learning experience is maximized consists of two major components, i.e. the entrepreneurship education and the enterprise education. Entrepreneurship education refers to a curriculum designed to embed the knowledge based education and theoretical context of entrepreneurship (established knowledge and theories) in the artificial environment (classroom context). Students are exposed to entrepreneurship knowledge, such as entrepreneurial behaviour, entrepreneur mind set, attributes and entrepreneur characteristics, preparation of business plans, management of small businesses, entrepreneurship process among others. For this component, the teaching pedagogy suggested include lectures, workshops, simulations, case studies, role play and guest speakers to name a few.

The second component, i.e. Enterprise education refers to a curriculum that emphasizes on fostering and enhances students' tacit knowledge (learning for entrepreneurship in real environment or setting). The curriculum provides an opportunity for students to actually execute their proposed projects proposed and learn through observation, experience and/or discovery. This educational approach grants an opportunity for students to learn and make decisions. Students are exposed to take responsibility and risk assessment on every action taken. Implicitly, learning by experience would boost their confidence level and practice the prudent concept in their work despite of the outcome of their decision (failure or success). New knowledge gained from this process which known as tacit knowledge is very valuable and would be useful in future.

Table 5. The difference between entrepreneurship education and enterprise education

Entrepreneurship Education	Enterprise Education
Emphasise on the delivery of existing entrepreneurship theories and knowledge	Emphasise on the cultivation of tacit knowledge through actual experience and business practices in self-selected project
Artificial setting or environment	Real setting or environment
Knowledge is learned	Knowledge is practiced
Instructors may non-practitioners	Learning through observation, experience and discovery, guided by practitioners
Only able to create interest and understanding	Able to create entrepreneurs among graduates
Hardly to measure the student's competency level as its subjective nature	The level of competency can be objectively measured
Pedagogy–lectures and tutorial, case study, simulation, role play and invited speakers.	Pedagogy–enterprise learning, entrepreneurship week, field work, student mall and business incubator.

6.1 The Rationale for the Proposed 4e Model

The 4e Model is the combination of theoretical and experiential learning. The traditional entrepreneurship education alone is less effective to develop entrepreneurial competencies. The traditional way of teaching only fit when the audience is attracted on the knowledge, but do not have the opportunity to put into practice. No doubt entrepreneurship education is important, however, without enterprise education which focus on learning through experience, the objective to develop enterprising graduates is amidst.

The 4e Model clearly categorises the contents based on level of study for both entrepreneurship and enterprise education. In general, the level of study is divided into four phases. At the beginning level, the curriculum is focusing more on entrepreneurship education and decreasing along the period of study. Subsequently, towards the end of the study, students are exposed more to enterprise education to allow for implementation of the business plan which was prepared in the earlier level.

The success of the 4e Model very much depends on the quality of input and methodology applied in the teaching and learning process. The inputs among others include infrastructure, lecturer/instructor, administrative support, curriculum and students. The 4e Model applies outcome based learning approach where the learning outcomes and entrepreneurial competencies can be measured objectively. The objective measurement made is based on evidence-based learning as compared to psychometric test which is too subjective. The 4e Model requires students to keep evidence into a personal folio, then evaluated by the academic advisor at the end of the semester. For students, the folio is a conclusive evidence for involvement in entrepreneurial activities during their study and would be very helpful during job interviews. By using the outcome based learning, the assessment of the goal achievement is far more credible and reliable

The 4e Model outlined the characteristic of the entrepreneurial graduates. The dispositional characteristics among others, are leadership, creative, innovative, dare to take calculated risks, persuasive, opportunity driven, integrity and with excellent communication skills. The model would also help to develop self-confidence among students by nurturing internal locus of control and positive affectivity. Preparing graduates with these attributes would enhance their entrepreneurial ability and competency. The 4e Model does not propagate creating a new venture is a must for graduates, but offer options and alternative routes for them to make a choice. Graduates may decide later whether to be self-employed, engage in the family business, or work with private or public sector or with non-governmental organisations (NGOs). Regardless of their choice of career, the most important is that graduates are able to apply and implement their entrepreneurial knowledge and skills in their new working environment thus benefiting the organization and society.

7. CONCLUSION

This paper provides an overview of the implementation of entrepreneurship education in all local IHLs. The information gathered is of paramount importance in reviewing the existing education model that leads to the proposal of new comprehensive entrepreneurship education model to be implemented by Malaysia's IHLs. The success of the proposed 4e Model depends crucially on the commitment and seriousness of the stakeholders involved. It should be noted that among the critical success factors for the 4e Model are (1) a conducive entrepreneurship ecosystem in IHLs, (2) strong support from the IHLs' management and stakeholders, (3) a clear policy on the entrepreneurial activities in IHLs, (4) commitment from members of faculty, departments, and units, and (5) an entrepreneur-friendly support system. Finally, the 4e Model is set to act as a guideline to inculcate entrepreneurial characteristics among graduates. These are the graduates who are opportunity seekers and job creators. As a conclusion, the successful implementation of the proposed model would lead to public IHLs being recognised as the best place to generate entrepreneurial human capital for the country.

REFERENCES

1. Cheng, M. Y., Chan, W. S., & Mahmood, A. (2009). The effectiveness of entrepreneurship education in Malaysia. Education + Training, 1(7), 555-566.
2. Dahlan, I., Hakimin, Y., & Azwin, N. (Eds.). (2010). The relationship between tacit knowledge and entrepreneurial

orientation among university students (Vol. 1). Kelantan: Universiti Malaysia Kelantan.

3. Dhliwayo, S. (2008). Experiential learning in entrepreneurship education. A prospective model for South Africal tertiary institutions. Education + Training, 50(4), 329-340.

4. Drucker, F. P. (1994). Innovation and entrepreneurship. Butterworth-Heinemann, Oxford.

5. Fayolle, A., & Gailly, B. (2008). From craft to science: Teaching models and learning processes in entrepreneurship education. Journal of European Industrial Training, 32(7), 569-593.

6. Gorman, G., Hanlon, D., & King, W. (1997). Some research perspective on entrepreneurship education, enterprise education and education for small business management: A ten year literature review. International Small Business Journal, 15(3), 56-77.

7. Jones, C., & English, J. (2004). A contemporary approach to entrepreneurship education. Education + Training, 46(8/9), 416-423.

8. National SME Development Council. (2007). SME Annual Report 2007. Kuala Lumpur.

9. Nurmi, P., & Paasio, K. (2007). Entrepreneurship in Finnish universities. Education + Training, 49(1), 56-66.

10. Pretorius, M., Nieman, G., & van Vuuren, J. (2005). Critical evaluation of two models for entrepreneurial education: An improved model through integration. International Journal of Educational Management, 19(5), 413-427.

11. Tan, S. S., & Ng, F. C. K. (2006). A problem-based learning approach to entrepreneurship education. Education + Training, 48(6), 416-428.

12. Van Vuuren, J., & Nieman, G. (1999). Entrepreneurship education and training: A model for syllabi/curriculum development. Paper presented at the 44th ICSB World Conference Proceedings.

10
CHAPTER

Cultural Heritage and the Public Domain

Bas Savenije, Annemarie Beunen

National Library of the Netherlands, NL

ABSTRACT

The recent report of the "Comité des Sages" recommends that "cultural institutions should make public domain material digitised with public funding as widely available as possible for access and re-use". One of the objectives of the Berlin Declaration on Open Access to Knowledge in the Sciences and Humanities is "encouraging the holders of cultural heritage to support open access by providing their resources on the Internet."

Libraries often are the only source for public domain material such as unique manuscripts. This position puts them in power when determining the conditions under which reproductions can be delivered. This position is prone to change as soon as public domain material is available via internet and thus can be copied by anyone.

We can observe a variety in re-use policies among cultural heritage institutions, in which not only libraries but also archives and museums are involved. And there certainly is no unanimity when it comes to commercial re-use. The situation becomes even more complicated when public-private partnerships are involved in which the commercial party poses restrictions on access and/or re-use.

The paper analyses the legal issues that are at stake in deciding about the library's re-use policy of digitised heritage material within the public domain. It also gives an overview of arguments pro and con open access without any restrictions. Its conclusion is in favour of no limitations for re-use, commercial or not.

Finally, it analyses public-private partnerships in the light of these conclusions.

Key words: cultural heritage, open access, digitisation.

INTRODUCTION

For centuries, libraries, archives and museums from across Europe have been the custodians of our rich and diverse cultural heritage. They have preserved and provided access to the testimonies of knowledge, beauty and imagination, such as sculptures, paintings, music and literature. The new information technologies have created unbelievable opportunities to make this common heritage more accessible for all.

Recently, the European Commission commissioned a 'Comité des Sages' to make recommendations on ways and means to make Europe's cultural heritage and creativity available on the Internet and to preserve it for future generations. In the United States the Association of Research Libraries (ARL) endorsed a number of principle recommendations to its members regarding the digitisation of cultural heritage. Both the Comité des Sages and the ARL emphasize the added value of digitisation. The Comité underlines that the digitised material can in itself be a driver of innovation and can be at the basis of new services in sectors such as tourism and learning (Comité des Sages, 2011) and the ARL stresses the added value for researchers (ARL Principles July, 2010). For over a century, libraries have participated in successful resource sharing cooperatives that have made content widely accessible. According to both the ARL and the Comité, the same spirit should govern commercial digitisation activities. In the best of all possible worlds, there would in our view be some level of free access to all content, with only special value-added services restricted to a subscription model.

A landmark in the discussion about Open Access to information is the Berlin Declaration on Open Access to Knowledge in the Sciences and Humanities. Referring to this Declaration, people often put emphasis on recent research publications. But the following is also one of the objectives of the Declaration: "encouraging the holders of cultural heritage to support open access by providing their resources on the Internet" (Berlin Declaration, 2003). Therefore, in the spirit of the Berlin Declaration, the ARL encourages its members' libraries to grant all non-commercial users "a free, irrevocable, worldwide, right of access to, and a license to copy, use, distribute, transmit and display the work publicly and to make and distribute derivative works, in any digital medium for any responsible purpose, subject to proper attribution of authorship". And: "If fees are to be assessed for the use of digitised public domain works, those fees should only apply to commercial uses" (ARL Principles July, 2010).

In our view, cultural heritage institutions should make public domain material digitised with public funding as widely available as possible for access and reuse. The public sector has the primary responsibility to fund digitisation. The involvement of private partners, however, is encouraged by ARL as well as the Comité des Sages. Private funding for digitisation is a complement to the necessary public investment, especially in times of economic crisis, but should not be seen as a substitute for public funding. As we can see from these reports there are a number of arguments in favour of digitisation and also of providing maximum accessibility to the digitised cultural heritage.

In this paper we will investigate the legal aspects of digitisation of cultural heritage, especially public domain material. On the basis of these we will make an inventory of policy considerations regarding reuse. Furthermore, we will describe the conclusions the National Library of the Netherlands (hereafter: KB) has formulated and the arguments that support these. In this context we will review public-private partnerships and also the policy of the KB. We will conclude with recommendations for cultural heritage institutions concerning a reuse policy for digitised public domain material.

LEGAL ASPECTS

Works in the Public Domain

An important legal principle is that intellectual property rights are monopolies of restricted duration (although some, like trademarks, can be renewed). At some point in time protection expires and that is a great asset because eternal protection would disproportionately hamper the users' freedom of action.

Within the whole European Union, copyright (leaving the neighbouring or related rights aside) expires 70 years after the death of the author. Where copyright in a work has expired or its maker has waived his or her copyright, the work is said to have fallen in the Public Domain.

The question whether the scanning of Public Domain material creates new copyright in the resulting scans is an interesting one. There is no clarity regarding this issue, however, either in the law or in jurisprudence. Two contrary views exist.

1. Digitisation does not create new copyright because it does not add any originality to the scanned work. It can be compared with making a photocopy of a book: the copier does not create copyright in the copies made. Scans too are merely faithful reproductions of the original. The same has been argued for photographs of two-dimensional paintings, prints or manuscript illustrations (Deazley, 2001; Beunen and Aalberts, 2002; Beunen, 2008; Chavannes, 2008, contra: Garnett, 2000; Stokes, 2001). Europeana's Public Domain Charter also states that scanned Public Domain works remain free of copyright; "What is in the Public Domain needs to remain in the Public Domain: Exclusive control over Public Domain works cannot be reestablished by claiming exclusive rights in technical reproductions of the works, or by using technical and or contractual measures to limit access to technical reproductions of such works. Works that are in the Public Domain in analogue form continue to be in the Public Domain once they have been digitised" (Europeana Public Domain Charter, 2010). The Comité des Sages is also of the view that simple digitisation should not give rise to new rights. Moreover, the European Commission has adopted this view in its 2011 Recommendation on the digitisation and online accessibility of cultural material and digital preservation. Here, the European Commission recommends that Member States ensure that material in the public domain remains in the public domain after digitisation.

2. Digitisation does create new copyright because creativity is added to the Public Domain work in the form of OCR modification, addition of metadata by experts etc. However, it goes without

saying that the bare source text without additions remains free of copyright (e.g. an annotated play by Shakespeare: his text without annotations remains in the Public Domain).

If a court were to rule in favour of the second view then the organisation that made the scans would still have the choice whether or not to claim copyright in its scans. A point of attention in this regard is the situation in which scanning is outsourced: Scanning companies sometimes claim copyright in their scans and it is advisable for libraries and other cultural heritage institutions to exclude this eventuality in the scanning contracts.

Database right

Producers of databases based in Europe have acquired their own form of protection, called the sui generis right or database right, by way of the 1996 European Database Directive. Whoever invests substantially in the obtaining, presentation or verification of the content of a database (which includes a website) acquires database right over the entire content. It makes no difference whether the content consists of Public Domain works or works in copyright. Database right is without prejudice to other parties' copyright in individual works, but it is an extra right that is created over the entire content of the database (Beunen, 2007). Thus, cultural heritage institutions can enjoy database right in databases and websites with Public Domain or other material. To secure the database right, it is wise for cultural heritage institutions to have scanning firms transfer their database right in the scanning contract.

The database right protects the investor against reuse by third parties of the entire database or substantial parts of it, without his/her permission. If a party reuses a non-substantial part, such as an individual Public Domain work, the database producer cannot prevent this on the grounds of his/her database right, even if it is for commercial purposes (although repeated and systematic reuse of insubstantial parts can amount to infringement).

Terms of use on the website

If the cultural heritage institution is not entitled to rights in scans of individual Public Domain works, it is doubtful whether the institution can prohibit reuse (commercial or otherwise) of scans from its website by means of terms of use. Existing Dutch case law is not unanimous on whether rights may be claimed to which one is not entitled on the basis of the law (Guibault, 2006). If a scanned work is indeed free of copyright, then its reuse can arguably not be constrained by terms of use. It is not uncommon for cultural heritage institutions to set terms of use prohibiting commercial reuse without permission. Generally users do abide by these, but if the scanned material is free of rights it is unlikely that these terms could be legally enforced. Furthermore, if a heritage institution receives an order for scans and agrees terms of use with the customer, these conditions would only apply to the customer in question. If the institution cannot claim rights over the scans supplied, then everyone is free to copy them from the customer's publication.

As it is not currently possible to establish with certainty whether cultural heritage institutions (and/or scanning companies) can claim rights in scans of Public Domain works, technical protection measures are sometimes chosen. This is more effective in practice than trusting to whatever protection offered through intellectual property rights or terms of use. The most extreme measure is protection against copying. However, from the users' point of view this is undesirable, because thus a user cannot make a copy for his/her private use or study, whereas copyright law allows this even for copyrighted works. A less drastic alternative may be found in the museum sector: many museums place only low resolution images online to prevent commercial usage. Commercial parties must order high resolution versions whereas the low resolution images are good enough for users who wish to make a copy for private use or study. In the case of texts this is not an interesting option, however, as high resolution does not yield any added value (with the exception of illuminated books of hours and the like).

It is also possible to insert a non-intrusive logo or watermark of the institution in self-scanned material, under or above the text. In this way the institution can, in the first place, make clear that it has financed the digitisation itself, thus generating goodwill: it is clear to whom the credit belongs. Secondly, the institution can then supply scans without the logo to commercial parties by means of its delivery service.

Policy Considerations

Providing access to cultural heritage is a task that serves the public interest. Cultural heritage institutions participate in successful resource sharing cooperatives that make content widely available. Providing access free of charge to Public Domain works seems an obvious course. The question we must address here is: do we wish to make Public Domain works online also freely available for every form of reuse?

The Public Domain Charter of Europeana and the Public Domain Mark, recently introduced by Creative Commons, argue in favour of such general free reuse. The principle of the Public Domain has been extended to digitised Public Domain works: a work that is free of copyright in its physical form should remain so in digital form. The Comité des Sages advocates for Public Domain works scanned through public funds: free access, free non-commercial reuse and the widest possible reuse possibilities for commercial parties.

Choosing to make self-scanned Public Domain works available online free for any reuse is a fundamental decision. It is irrevocable and will apply to all online types of Public Domain content that a heritage institution places or intends to place online. Everyone, including commercial parties, may then freely download and reuse this content digitised with public money.

Arguments for free reusability of online Public Domain works for every type of reuse The public funding of cultural heritage institutions implies that free reusability is logical. For example, libraries carry out their tasks mainly with public money. This applies both to the acquisition of the collection and the execution of digitisation projects. The Public Domain principle argues in favour of totally free reusability: as soon as all rights have expired, reuse and modification of the material online must be free of charge and without any restriction. Copyright already lasts for a very long time (according to many, far too long) and after it has expired all practical and financial use impediments must be lifted. In comparison: the Open Access movement argues that information that has been created with public funds must be made freely accessible to the taxpayer. The definition of Open Access in the Berlin Declaration encompasses in addition to free access also free reuse 'for any responsible purpose'; this includes in any case non-commercial reuse and according to some also commercial reuse. Incidentally: if a customer orders scans/reproductions, for which the institution must perform activities and therefore incurs costs, it is logical that these costs are passed on to the customer. If one adheres to the Public Domain principle then — in the case of Public Domain materials — nothing more than a cost-recovering fee should be charged for this service, both for non-commercial and commercial use.

Totally free reusability is advantageous for all types of users and promotes usage. Wider online distribution may be expected and wider brand exposure if scans carry the logo of the institution. Cultural heritage institutions may also regard it as their task to keep Public Domain materials freely accessible and available for use to counter the restrictive terms of use imposed by companies that exploit such content via Public-Private Partnerships with the heritage institutions.

It is an illusion to think that it is possible to maintain control over content on the Internet and over the quality and context of its reuse. The KB, for instance, currently relies on the goodwill of commercial users to request permission, but such permission is not legally required if the KB cannot claim rights in the scans of Public Domain works. Free reusability also avoids the problem of taking measures against commercial 'abuse': taking legal action in retrospect is not possible if the library cannot claim any rights in the scans.

With low resolution scans online, the demand for orders continues to exist: commercial parties wishing to receive high-resolution materials (and without watermark) in large bulk may continue to request these against payment

Disadvantages of free reusability of online Public Domain works for every type of reuse

Because the choice for free reusability is a fundamental one, it is irrevocable and will apply to all Public Domain types of material scanned with public funds that the heritage institution chooses to put online.

The decision to make content freely reusable is not necessarily a financially viable one. It might destroy a potential source of substantial income.

Making Public Domain content that has been placed online freely reusable gives rise in a way to arbitrariness. This is because whether a Public Domain work is freely available for reuse would depend on whether the library has decided to scan it and place it online. Other Public Domain material could still only be ordered for a fee.

In the case of free reusability, a distinction needs to be made between Public Domain material scanned with private or with public funds. However, this distinction is not legally relevant: the free reusability of self-scanned material may cause users to also question the use restrictions imposed on content scanned by Public-Private Partnerships. These restrictions are legally untenable if no intellectual property rights rest in the scans (the database right only offers protection against reuse of substantial parts of the database). This is something users can already work out for themselves, but a clear distinction (Public Domain material with/without Public Domain Mark) will make this evident sooner.

Commercial reuse

Cultural heritage institutions need to find their way amid the tension between on the one hand facilitating optimal access and usage, and on the other developing cultural entrepreneurship (generating their own income). An alternative to free reusability could be a policy based on the principle of profit: charging costs for commercial reuse of digitised Public Domain works, whether they are online or not. The access to Public Domain material that the institution has digitised itself would then be free for everyone, as would reuse for personal study, education and research. However, for commercial reuse of this material prior permission would always be required and a fee would be charged.

Not making content freely available for commercial use has a number of advantages. It creates the possibility to monitor the quality of commercial reuse in advance and therefore prevents such reuse in a quality and/or context with which the library does not wish to associate itself. Furthermore, a low price threshold for commercial reuse is acceptable and offers the possibility of extra income, for instance by negotiating a beneficial deal for a library through packaging. In addition, as more Public Domain content is placed online, more downloads occur. If commercial parties themselves download the content to be reused, a cultural heritage institution will receive income without incurring additional costs for delivery of the content. Finally, in the perception of (a part of) the general public the content will acquire more value: "whatever is free is without value", is an argument that still regularly plays a role (wrongly, as it happens) in Open Access debates.

Public-Private Partnerships

Digitisation is expensive and cultural heritage institutions are not always able to raise, in part or in total, the necessary funds themselves. Often they are dependent on their national government. The European Commission does not subsidise, or only to a limited extent, the digitisation of existing collections. The Comité des Sages too regards this, within the European Union, as primarily the responsibility of the Member States: "Member States will need to considerably step up their investments in digitisation" (Comité des Sages, 2011, p. 6). National governments have sometimes created programmes for financing digitisation activities, but these are under a great deal of pressure because of the economic crisis.

In such a situation, cooperation with a commercial party that can bear the costs of scanning can offer a solution. But such a deal often has a negative side: in order to earn back its investment a company usually negotiates some form of exclusivity in order to be able to exploit the scanned heritage commercially, for instance via subscriptions for libraries. This exclusivity can be agreed for a temporary embargo period, but embargos for an unlimited period also occur. This results in a temporary or permanent 'digital lock-up' of the own collection: although the cultural heritage institution receives its own digital copy, it is not (yet) allowed to make this available to the public, even though it often contains Public Domain content that — without these conditions of exclusivity — could have been placed freely online, unrestricted by any intellectual property rights. Unfortunately, this is precisely the reason why companies opt for Public Domain content when they embark on a cooperation with cultural heritage institutions: in this way their commercial exploitation of the cultural heritage will not be hindered by the copyright of other parties. The principle of freedom of contract enables them to subsequently negotiate exclusive exploitation in the contract with the cultural heritage institution (Beunen, 2010).

As noted above, financial considerations are usually the decisive factor for accepting this 'digital lock up': another party pays for the costly digitisation and in addition to scans the institution sometimes receives a percentage of the exploitation income and/or (temporarily) a free subscription to the scanned content.

The Comité des Sages has recently stressed again that Public Domain content in the analogue world should remain in the Public Domain in the digital environment. The exclusivity which the private partner may stipulate is therefore an important disadvantage of public-private cooperation. The cultural heritage institution does get its scans, but is not yet allowed to make them available to the public because the company wishes to recoup its scanning investment by means of commercial exploitation. It is important to carefully consider how long one wishes to accept this 'digital lock-up'. A limited embargo period would seem to be the obvious route, as indeed the Comité des Sages argues. The Comité is of the view that an embargo period up to a maximum of 7 years is acceptable, which was also recommended by the European Commission in its 2011 Recommendation.

In practice, other alternatives may also be negotiated, such as free access for certain target groups or national IP numbers, so that the company can only exploit the material commercially abroad. On the other hand, such an alternative prevents aggregation by the European cultural heritage portal, Europeana. This is also the reason why the European Commission is opposed to the 'exchange' of a (longer) embargo period for access in the home country. Then again, the funding for the acquisition and/or maintaining of cultural heritage collections comes from national means: this makes it unacceptable for Public Domain material to be digitally accessible to national users only for a fee.

Viewed from a long-term perspective Public-Private Partnership are, nevertheless, certainly attractive for the institutions involved. On the one hand, the funds needed to scan materials oneself are not available and one is faced with a choice: not to have them scanned or to do via a Public-Private Partnership, where in the case of a temporary embargo the access restriction expires after a number of years. In other words: there is either no digital availability at all or a temporary, restricted one; one way or another, the user will eventually benefit from the latter option. For cultural heritage institutions whose task it is to preserve their collection, if possible 'for eternity', this argument applies all the more strongly.

They must, however, be alert to the fact that scans that are made by commercial parties do not always meet the quality requirements for digital preservation. Digitised material should preferably still be accessible in a hundred years' time. This is not a primary concern for the companies, but should be for the cultural heritage institutions. Therefore, it has to be preferred that the company involved does not get exclusive permission to scan material (or only for a limited embargo period), otherwise the possibility of self-scanning (to produce better quality scans) or having scans made by other parties is completely excluded. It is also preferable if the scan activities are restricted to a clearly specified form of exploitation: granting carte blanche permission for any kind of (commercial) future form of exploitation is certainly not desirable. Finally, agreements must be made about what end users may do with the material that the company wishes to scan and exploit. Furthermore, it is not advisable to give the company permission to itself supply the digitised material (whether or not for a fee) to other parties for the purposes of reuse.

It is essential that the cultural heritage institution agrees with the selection of material to be scanned so that the company does not have a completely free hand in its choice of materials. In the case that access is restricted to national users only it is possible to negotiate that a certain percentage of the collections is made accessible worldwide. This may apply permanently to a fixed part of the collection, whereas for a different percentage of the material accessibility may vary for the benefit of, for instance, web exhibitions.

The Comité des Sages rightly points out that Public-Private Partnership agreements must be transparent and concluded by means of a written contract. So-called non-disclosure clauses are not acceptable. They prevent the terms from being subject to public accountability

Policy of the KB

With the help of the above analysis and working within the framework of its recent strategic plan, the KB has formulated a policy for access to and reuse of Public Domain material.

The core business of the KB is to offer permanent access to everything that is published in and on the Netherlands. The most important policy principles for the KB are:

- Increase use of the collection by transitioning as quickly as possible to a digital collection and digital services. This implies acquisition of born digital material and digitisation of the printed collection.
- Increase use of the collection by cooperation with relevant partners in the information sector who have a direct relationship with the target groups of the KB; these include other research libraries and public libraries. In this regard, the KB also wishes to strengthen the national infrastructure.
- Long-term preservation of the collection; in the case of digital storage, in cooperation with the other cultural heritage institutions within the framework of the National Coalition for Digital Preservation.

Reuse: Non-commercial and commercial

Two considerations are leading with regard to access and reuse of Public Domain material.

1. The KB is financed with public money and plays an essential part in making Dutch cultural heritage available, in cooperation with other cultural heritage institutions. This implies that the principle 'Public investment implies public benefits' has high priority for the KB.

2. The aim of increasing usage implies minimal impediments to that use, not only optimal technical accessibility but also as few financial or supervisory barriers as possible.

The conclusion is that the KB has decided not to impose any restrictions on non-commercial reuse of Public Domain material online scanned by public money. Moreover, the KB chooses for free reusability for any form of reuse, including commercial use. The KB does not regard commercial activities by private parties or other external initiatives as competition. If an external institution supplies added value to the freely accessible scans, the market will make clear whether users wish to pay for that. They will in that case pay for this added value; the Public Domain documents themselves will remain freely accessible via the website of the KB.

The KB's view is that it does not itself wish to undertake commercial activities with Public Domain material. Both the collection and the added value supplied by the KB with public money must be accessible to the public without cost. An exception is made at the moment for facsimile publications by third parties of valuable parts of the special collections of the KB. Such an activity does not interfere with the accessibility via the Internet, but is an additional activity that, if there is a demand for it, may be carried out for more than a cost-covering fee.

The KB is strongly against imposing restrictions that are underpinned by the argument that the future may bring new possibilities for a business of its own. Retaining to use restrictions only on the basis of the 'just in case' argument very often appears to be a serious obstacle to innovation.

The argument that the KB via use restrictions could monitor the quality of the reuse is not a convincing one. The KB is of the view that it is undesirable to monitor how customers use its collection or to formulate an opinion about the way the collection is used. This applies both to commercial and non-commercial use.

Nor does the KB share the objection that the public would ascribe a higher value to a collection if it is not made available free of charge. This argument is regularly used with respect to the Open Access movement: quality, however, is not determined by price but by the value of the content, which can be measured in a different way than a price tag.

The above considerations have led to the conclusion that the KB will not impose any restrictions on the commercial reuse of the Public Domain material which it scanned with public money. To make it clear that every reuse is allowed the KB intends to put this material online with the Public Domain Mark issued by Creative Commons. This organisation provides a standardised system of easily understandable terms of use. The PD Mark fits our purpose because it leaves any reuse free (www.creativecommons.org/about/ pdm). The other available Creative Commons licences cannot be used for PD material since they can only be applied by rights holders to works in which copyright subsists.

Public-Private Partnerships

The KB has also concluded Public-Private Partnerships with regard to Public Domain material. The KB has insufficient funds to digitise its entire collection itself. There is a national programme called Metamorfoze for the digitisation of printed cultural heritage (www.metamorfoze.nl), but the emphasis here is on conservation. This means that the most vulnerable material takes priority and that is not, in most cases, the material prior to 1800.

In the first place, the KB has concluded a contract with Google for Public Domain material (1700–1870). The scanning is done on a non-exclusive basis, so the library is free to have the same material scanned elsewhere. The selection is carried out by staff of the KB. Google does not charge a fee for access to the books it has scanned. The cooperation with Google is subject to a number of conditions. Books that are in the Public Domain in their printed form must remain in the Public Domain in digital form as well. This is achieved due, among other things, to the fact that the KB receives a copy of the scans, which it can make available through its own website as well as through Europeana. In this way the KB has also ensured that it can take responsibility for longterm storage, as Google does not offer any guarantees in this regard. This also means that users, who for reasons of privacy do not wish to access the scans via Google, have alternatives. Neither the KB, nor a third party are allowed to use the Google scans for commercial activities (which touches upon the issue of enforceability discussed in paragraph 2). Commercial activities are permitted if the KB or a third party makes its own scans. In other words: a certain exclusivity adheres to use of the Google scans, but not to the original books or other scans that are made of them. No non-disclosure has been agreed with Google, which means that the KB can offer public accountability for the content of the contract.

A second form of Public-Private Partnership has been concluded between the KB and ProQuest for our special Public Domain material (1450–1700), within the framework of the ProQuest programme Early European Books Online. ProQuest recoups the costs of the digitisation by means of a fee for access.

The KB has agreed the following with ProQuest:

- The KB receives a copy of the scans for long-term storage.

- A temporary embargo, after which the material is made available for free in Open Access, via among others the KB website.
- Complete access for users in the Netherlands during the embargo period.
- The KB may also offer permanent international access to a percentage of the collection (selected by the KB). In this way the KB can set up virtual exhibits about its top pieces.
- The KB may offer temporary international access to a percentage of the collection that may vary (selected by the KB), also for the benefit of temporary virtual exhibitions.
- A limited non-disclosure clause is agreed with ProQuest which allows the KB sufficient accountability.

An evident disadvantage of this cooperation is the financial barrier which foreign users experience when they access the collection. The alternative would, however, have been that the material could not have been digitised for the coming years. That is why the KB opted to make access to the Dutch target group a priority. This is the majority of the anticipated users. They are, moreover, the users who should profit the most from the adage 'Public investment implies public benefits', since the acquisition and maintenance of the collection have been paid with Dutch taxpayers' money.

CONCLUSION

Libraries and other cultural heritage institutions develop collections and maintain them for the most part with public money. They make the physical parts of collections that fall within the Public Domain available without financial restrictions.

Collections which fall within the Public Domain in physical form should fall within the Public Domain in digitised form as well, certainly if they have been digitised with public funds. Public investment implies public benefits.

The law does not make clear whether a cultural heritage institution is entitled to copyright in scans made of Public Domain material. But even if this were the case the institution should make the material available free of restrictions for both non-commercial and commercial reuse, if the Public Domain material was digitised with public money. Value added by a commercial party justifies income for this party provided that the scans themselves remain freely accessible without restrictions at the heritage institution's website.

Public-Private Partnerships can offer an institution the necessary financial means for digitisation in order to increase usage of the collection. Some forms of Public-Private Partnerships entail a temporary embargo on free access. In light of the awareness that without such an agreement digitisation would be impossible, such temporary embargoes are acceptable. In the long term, permanent free access is realised in this way. By means of supplementary conditions (free national access, exceptions to the embargo) the advantages of such an agreement may outweigh the evident disadvantages

REFERENCES

1. Association of Research Libraries (ARL). Principles to Guide Vendor/Publisher Relations in Large-Scale Digitization Projects of Special Collections Materials, endorsed 26 July 2010. Retrieved June 8, 2012

2. Berlin Declaration on Open Access to Knowledge in the Sciences and Humanities (2003). Retrieved June 8, 2012 from http://oa.mpg.de/lang/en-uk/berlin-prozess/berliner-erklarung.

3. Beunen, A.C. and B.P. Aalberts (2002). Exploiting Museum Images. In R. Towse (Ed.), Copyright in the Cultural Industries (pp. 221–232). Cheltenham: Edward Elgar Publishing.

4. Beunen, A.C. (2007). Protection for databases: the European Database Directive and its effects in the Netherlands, France and the United Kingdom. Nijmegen: Wolf Legal Publishers. (Ph.D. thesis Leiden University) Retrieved June 8, 2012 from https://openaccess.leidenuniv.nl/handle/1887/12038.

5. Beunen, A.C. (2008) Musea als makelaars in beeldmateriaal. Maandblad voor Vermogensrecht 2008, 7/8, 163–167. Retrieved June 8, 2012

6. Beunen, A.C. (2010). Erfgoedinstellingen online: musea, bibliotheken en archieven in de praktijk. In L.P.C. de Belder (ed.), Documentatie van Cultureel Erfgoed in juridisch perspectief (pp. 21–45). Amsterdam: deLex. Retrieved June 8, 2012 from https://openaccess.leidenuniv.nl/handle/1887/16359.

7. Chavannes, R. (2008). De kunst van het onzichtbaar blijven. Het auteursrecht van de reproductiefotograaf. In N. van Eijk and B. Hugenholtz (eds.), Dommering-bundel, Opstellen over informatierecht aangeboden aan prof. mr. E.J. Dommering (pp. 29–44). Amsterdam: Otto Cramwinckel. Retrieved June 8, 2012 from http://www.chavannes. net/assets/Onzichtbaar.pdf.

8. Commission Recommendation on the digitisation and online accessibility of cultural material and digital preservation (2011, October 29). Official Journal of the European Union, L283, 39–45. Retrieved June 8, 2012 from http://eur-lex.europa.eu/LexUriServ/LexUriServ.do?uri=OJ:L:2011:283:0039:0045:EN:PDF

9. Deazley, R. (2001). Photographing Paintings in the Public Domain: A Response to Garnett, European Intellectual Property Review, 23, 179–184.

10. Directive 96/9/EC of the European Parliament and of the Council of 11 March 1996 on the legal protection of databases, Official Journal L 077 (1996, March 27), 20–28..

11. Europeana Public Domain Charter (2010). Retrieved June 8, 2012.

12. Garnett, K. (2000). Copyright in Photographs. The Problem of Photographs of other Works, European Intellectual Property Review, 22, 229–237.

13. Guibault, L. (2006). Wrapping Information in Contract: How Does it Affect the Public Domain? In L. Guibault and P.B. Hugenholtz (eds.), The Future of the Public Domain - Identifying the Commons in Information La (pp. 87–104). The Hague: Kluwer Law International (Information Law Series 16). Retrieved June 8, 2012 from http://www.ivir.nl/publications/guibault/wrapping_information_in_contract.pdf.

14. Proposal for a Directive of the European Parliament and the Council, Amending Directive 2003/98/EC on re-use of public sector information (2011, December 12). COM(2011) 877 final. Retrieved June 8, 2012

15. Report of the 'Comité des Sages' (2011, January 10). The New Renaissance. Retrieved June 8, 2012

16. Stokes, S. (2001). Photographing Paintings in the Public Domain: A Response to Garnett, European Intellectual Property Review, 23, 354.

11

CHAPTER

Library Planning in Germany

Friedrich Geisselmann

K.G. Saur, Munich, printed in Germany

ABSTRACT

The position and the efficiency of German libraries are rather ambivalent. On the one hand the infrastructure is quite well developed: 2000 public library systems with full time personnel, 275 million books issued, 75 million Euro for acquisitions; 280 scientific general libraries, 60 million books issued, 1300 million Euro for acquisitions. Nevertheless there are a lot of problems, which arise from the administrative structure and the political position of libraries.

The Position Of Libraries In The Political Process

The position and the efficiency of German libraries are rather ambivalent. On the one hand the infrastructure is quite well developed: 2000 public library systems with full time personnel, 275 million books issued, 75 million Euro for acquisitions; 280 scientific general libraries, 60 million books issued, 1300 million Euro for acquisitions. [1] Nevertheless there are a lot of problems, which arise from the administrative structure and the political position of libraries.

Most scientific libraries are maintained by the states, and only very few by the federation. This means that library policy is the responsibility of the states. The federation has no competence for scientific libraries. Although there are some coordinating institutions, concepts for a nationwide library planning are rare. Public libraries are maintained by autonomous local government. There are no specific library laws, and political concepts in this field are rare as well. In the political process libraries are of rather little importance. This is also more and more the case with university libraries. University laws deal less and less with libraries. The state administration is losing its influence and coordinating power.

Nevertheless, libraries are influenced by the Federal Ministry for Education and Research (**B**undesministeriums für Bildung und Forschung, BMBF) and by the National Science Foundation (Deutsche Forschungsgemeinschaft, DFG), the main bodies that finance projects and cooperative plans. Both institutions are In fact involved in library planning.

The main problems are:

- How to get adequate finance for libraries?
- How to promote innovation either by advanced training of personnel, by exploitation of international best practice models, or by cooperation in new models of library service?
- How to find the new role for libraries in a digital world? A specific feature of German libraries is intensive cooperation:
- A system of distributed acquisition of all relevant scientific books financed by the DFG (Sondersammelgebietsplan: WEBIS).

- There is an extensive interlibrary loan and document delivery system (Subito)
- Cooperative cataloguing is used in all scientific libraries. Around the union catalogues service centres developed, assisting libraries in administration of local library systems, organizing central library services, and providing information for their libraries and in some cases even organizing regional consortia (Arbeitsgemeinschaft, 2003)
- In the last decade a lot of commissions on the national level existed and there was a central institute: German Libraries Institute (Deutsches Bibliotheksinstitut, DBI).

This institute, however, was dissolved in 1998 due to an evaluation it couldn't stand and due to political problems in cooperative federalism. This liquidation caused great problems. The library associations (DBV, VDB, BIB, BDB) were not able to prevent this.

New Organization For Cooperative Work

After the dissolution of the German Libraries Institute (DBI) there were many discussions to find ways to get out of this disastrous situation. Three different methods were attempted to re-establish library planning:

- Finding new ways of professional cooperation.
- Introducing a new planning process organised by the library organisations and the Bertelsmann Foundation.
- Using a combination of recommendations of independent institutions and projects to establish new services in the field of digital libraries.

The German Library Association (Deutscher Bibliotheksverband, DBV) is a centre of professional work. Commissions and expert groups always played an important role in German librarianship. The honorary work of experts distributed in libraries is part of the decentralized library system in Germany. Some librarians think this non-professional work is unprofessional, but experiences with professionals in central institutions show that the work of commissions and expert groups is indispensable. This work was organised by the German Libraries Institute (BDI) for a long time. It was part of its legal mandate. All librarians agreed that this work should continue. There was some discussion whether the association of the institutions or those of the professions (VDB and BIB) should organise this. It was decided that the association of the institutions (DBV) should assume this task. Three commissions were established for core activities of libraries:

- Commission on Management of Libraries.
- Commission of Library Services.
- Commission of Legal questions

These commissions are responsible for all groups of libraries. Expert groups on the other hand deal with problems relevant only for one or several groups of libraries, i.e. sections of the association, such as problems in acquisition for scientific libraries, school libraries, public libraries etc.

After the decision to dissolve the Deutsches Bibliotheksinstitut an intense and long struggle to set up a new institution began. I won't describe this discussion here in detail. Only the underlying goals are relevant, and not the problems which arose about financing such an institution.

The first step was the examination which tasks were indispensable on the national level. There was an agreement between the states and the federation about these tasks in 1999. The next step was a concept, a statute and a plan how to finance an Innovation Centre for Libraries (Innovationszentrum für Bibliotheken, IZB). This Institute was supposed to be set up by the Foundation of Prussian Heritage in Berlin. The main goal of it would be to encourage innovation in libraries. A minor goal would be to take over several services for libraries, especially German library statistics (Deutsche Bibliotheksstatistik,DBS). Although there was a common agreement between the federation and all ministries of culture and enough money was provided, the concept failed in the end. The reason was a general discussion about such institutions of cooperative federalism. At the moment there is a tendency to separate tasks of federation and states.

Instead of such a central institution the German Library Association (DBV) proposed a decentralised competence network for libraries (Kompetenznetzwerk für Bibliotheken). We realized that this would be the only way to save the goals of the innovation centre (IZB). Partners in this competence network will be:

- the 6 regional union catalogues,
- 5 libraries of national relevance (DDB, SBB, BSB, TIB, SUB Göttingen),
- German Library Association (DBV),
- Central institutions for public libraries (Fachstellenkonferenz),

- ekz bibliotheksservice GmbH .
  The network will bring together the competence of institutions already active in this field. There will be synergetic effects between the tasks of these institutions and new tasks. The mains goals will be:
- to collect information for planning and decisions for federal policy and for the states,
- to identify strategic priorities,
- to strengthen the role of German libraries on the international level and to encourage international contacts.

The next step in making the concept was to identify the tasks, which should be fulfilled by the network. An agreement was reached about this. About 22 additional positions for personnel will be necessary.

The main problem now is financing such new tasks. Extra funding is very difficult to get in Germany today. It is decided that such a competence network will be established starting 1.1.2004. Only parts of the program can be financed at the moment. It will be necessary to look for projects to get the program started with money for a limited time and by means of the participating institutions.

Bibliothek 2007 - A Strategic Planning Model For Librarty Development

National library plans were made several times before. In the seventies planning was widespread in all parts of administration. A typical product was the Library Plan 1973 (Bibliotheksplan 73). It was influenced by regional planning and created paradigms for types of libraries which should build up a system of libraries. This influenced political decisions in Germany for a long time. 'Bibliotheken 93' was a document of the next generation. Planning methods hadn't been successful. In 'Bibliotheken 93' the discussion was summed up and a collection of documents relevant for political decision was offered. In which way can a library plan work today? It will have to be made up in a different way than before. The problems in German librarianship are mostly those of getting plans realised, not of missing concepts. The main problem is the lack of political power, because politicians see little relevance in library problems. Therefore such a library plan has to be a kind of a political plan.

Due to this it was decided that the library associations together with the Bertelsmann Foundation should develop a new plan, Bibliothek 2007. The Bertelsmann Foundation provides financing of the project, and the Bundesvereinigung Deutscher Bibliotheksverbände (BDB) - an umbrella organisation of DBV, VDB, BIB, ekz, Bertelsmann Foundation and the libraries of the Goethe Institute - provides advice.

A few words about the Bertelsmann Foundation. The foundation is the largest owner of the publishing house Bertelsmann Verlag. It is an operative foundation; this means it doesn't simply provide money, but works on its projects. Bertelsmann Foundation is in fact a very important think tank in the field of culture, education and libraries. The foundation is active in a lot of projects in libraries, especially public libraries, but nowadays in scientific libraries as well. Some current projects are: Deutsche Internetbibliothek, a digital library for public libraries, BIX, a system of measurement of efficiency and a ranking of libraries; it is one of the partners maintaining the City Library of Gütersloh etc.

Bibliothek 2007 can be seen as a strategic planning model to influence the discussions of librarians, the public and politics about further development of library infrastructure and library services in a knowledge society. The model also deals with the optimal use of money, but the aim is not rationalization or saving money. The political process following the publication of the study is an important part of this project. Yet this part of the project is not defined at the moment. The year 2007 refers to the projected success of the discussion. The project started in 2002 and will be finished in 2003. The applied methods are:

- Qualitative interviews with representatives from different relevant groups. This was complemented by an analysis of opinion surveys in different types of libraries.

Infas, an institute of public opinion, did this.

- Surveys of international best practice models (Denmark, Finland, Singapore, UK, USA). The international surveys will show the political situation, the reasons of success for the library systems in the respective countries. This part is done by a management consultant (Booz Allen Hamilton). A working group of experts of all groups of libraries and the Foundation will have to work out the master plan.

Planning Of Digital Libraries

A result of Bibliothek 2007 is that libraries are acknowledged as an important place of social interaction. Libraries therefore shouldn't be mere virtual realities. Nevertheless, digital libraries are the most important problem in the near future. Digital libraries have the potential to improve and accelerate information for the user dramatically. On the other hand the role of all partners in the information chain (authors, publishers, book trade, libraries and users) have to be examined. There is a huge potential of development and a lot of work to do. Therefore planning is especially relevant for digital libraries. The main themes of planning are:

- The strategic concept for the developments in libraries.
- The structure of the new services: will the digital library be national, regional or local? Some people think the problems are too big for the average scientific library.

Digital documents can be stored on a central server. The average library should concentrate on its physical documents and offer access to a national digital library.

On the other hand most libraries think they should offer a portal to all information. This portal should offer material necessary for local users. As licenses have to be paid locally a local portal is the consequence.

- The problem of financing digital publications, especially e-journals.

The strategic concepts were dealt with in several recommendations by institutions established in the administration of education and science. The best paper is*Empfehlungen zur digitalen Informationsversorgung durch Hochschulbibliotheken*(Recommendations for digital information in university libraries, 2001) by the Council for Science. It deals with all problems from the users′ point of view. Another large-scale planning process was organised by the Federal Ministry of Education and Science. Several parallel groups worked on this field. A management consultant Arthur D. Little worked out the most ambitious recommendation (Zukunft, 2002). But the result was very poor and the statement of the Ministry was rather vague.

I mention here the recommendations of Conference of University Rectors (Hochschul Rektoren Konferenz, HRK) in the paper *Zur Neuausrichtung des Informations- und Publikationssystems der deutschen Hochschulen/Hochschulrektorenkonferenz* (2002), because these are very relevant for questions of library planning and correspond with the recommendations of the Conference of Ministries of Culture. The Conference of University Rectors recommended widespread consortia for electronic journals. There were some problems in using online information identified:

- scientists are sceptical about electronic media (this is different according to different disciplines),
- publishers rely on print editions,
- problems of long term archiving and accessibility are not solved,
- the possibility that electronic publication are not used enough. In this situation pay per view isn't the right option, because this will impede usage and the advantages of electronic publications will get lost. Therefore site licences for a great amount of publications are recommended, at least for some years. In the meantime detailed statistics of usage should show which journals are used. Upon these statistics new pricing models should be developed. To finance such a licensing policy a special programme financed by the federation should be implemented. The tax on digital publications should be the same as the tax on printed publications.

The situation in German libraries is as follows: consortia for all-inclusive licenses are widespread. For example the University Library of Regensburg has about 9200 print journals and licenses and access to free journals for about 9800 titles. A lot of these titles were added by cross-access e.g. 960 Elsevier titles in electronic form where we formerly had 220 Elsevier titles in print. A great number of these electronic titles are not used at all. We have a platform independent from publishers websites, the Elektronische Zeitschriftenbibliothek (EZB) and a good usage of these titles. The problem at the moment is that due to the financial crisis in Germany our budgets are cut down for 10%. We don't know how publishers will react to this.

The second recommendation of the Conference of University Rectors is the establishment of electronic publications by the universities themselves. These servers should comprise all scientific publications of members of the university and e-learning material. A system of peer reviewing should be added. University-owned publishing houses should be installed at least in some cases. New models of cooperation between such servers and publishers should be developed. Nobody objects to these recommendations. The problem is how to realize this and how to change the minds of the authors.

The Conference of Ministries of Culture, the ministries responsible for financing universities, recommended that the federation should pay for electronic publications. On the other hand the ministries expect that more money will strengthen publishers who are able to raise prices more and more. Therefore new pricing models should be discussed. So, there are two opposite points of view: libraries need more money to get broad access; on the other hand the finance has to be reduced to save money and to resist to the unfair praxis of some publishers. I'm afraid that this discussion will be decided by the Ministries of Finance, which cut down our budgets.

Projects are financed on the national and on the regional/ local level. As the federation has more money to spend national projects are highly relevant. The most relevant national projects are brought together in Vascoda, a portal that will be released on

IFLA in Berlin, on August 8th, 2003. This portal includes 4 'Informationsverbünde' (information networks) - networks connecting special libraries and information centres. The main business model is pay per view. These institutions are financed by the Federal Ministry for Education and Research (BMBF). On the other hand there are a lot of virtual special libraries (Virtuelle Fachbibliotheken). These collect free electronic material and build up union catalogues for special fields. These institutions are financed by the National Science Foundation (DFG). A special type of partner is EZB, which is a voluntary informal organisation run by one library. The aim is to integrate all these projects and to offer not only a general national portal but also national portals for all scientific disciplines. On the other hand there are a lot of local or regional projects. In all regions with union catalogues portals are already developed or under construction. For our theme is relevant the project Distributed journals server (Verteilter Zeitschriftenserver, VZS). It is planned to store e-journals on servers of the union catalogues. Thus the availability of journals already acquired can be ensured. The aim for the next year will be to combine all relevant projects under one roof and to harmonize the different solutions. This has to be done under the roof of Vascoda, which will develop as a national digital library.

To conclude: the situation in Germany is rather difficult due to problems in administration. Nevertheless libraries made substantial progress in the last years.

REFERENCES

1. *Arbeitsgemeinschaft der Verbundsysteme*. Frankfurt am Mainz : Sekretariat der AG Verbundsysteme, 2003

2. *Empfehlungen zur digitalen Informationsversorgung durch Hochschulbibliotheken*. Greifswald : Wissenschaftsrat. Drs. 4935/01. 13 Juli 2001.http://www.wissenschaftsrat .de/texte/4935-01.pdf

3. *Zukunft der wissenschaftlichen und technischen Information in Deutschland*. Schlussbericht. September 2002. http://www.dl-forum.de/Foren/Strategiekonzept /Schlussbericht.pdf

4. *Zur Neuausrichtung des Informations- und Publikationssystems der deutschen Hochschulen*. Hochschulrektorenkonferenz. Empfehlung des 198. Plenums vom 5. November 2002. http://www.hrk.de/ downloads/Empfehlung_Bibliothek.pdf

WEB SITES REFERRED TO IN THE TEXT

Die Arbeitsgemeinschaft der Verbundsysteme: http://www.ag-verbund.de

BDB - Bundesvereinigung Deutscher Bibliotheks verbände. http://www.bdb-dachverband.de/index2.html

Bertelsmann Foundation. http://www.bertelsmann-stiftung. de/index.cfm

BIB - Berufsverband Information Bibliothek. http://www.bdb-dachverband.de/englisch/index.html

Bibliothek 2007. http://www.bibliothek2007.de

BMBF - Bundesministeriums für Bildung und Forschung. http://www.bmbf.de/

DBS - Deutsche Bibliotheksstatistik.http://www.bibliotheksstatistik.de/

DBV - Deutscher Bibliotheksverband. http://www.bibliotheksverband.de

DFG - Deutsche Forschungsgemeinschaft. http://www.dfg.de/

Ekz : Service für Bibliotheken. http://www.ekz.de/

EZB - Elektronische Zeitschriftenbibliothek. http://rzblx1.uni-regensburg.de/ezeit/ezb.phtml

Goethe-Institut. http://www.goethe.de/

HBZ - Hochschulbibliothekszentrum des Landes Nordrhein-Westfalen.http://www.hbz-nrw.de/

HRK - Hochschulrektorenkonferenz. http://www.hrk.de/

Infas TTR. http://www.infas-ttr.de/

Subito. http://subito-doc.de

Vascoda. http://www.vascoda.de

VDB - Verein Deutscher Bibliothekare (Association of German Librarians).http://www.vdb-online.org/

WEBIS. http://webis.sub.uni-hamburg.de

NOTES

[1] Library statistics (Deutsche Bibliotheksstatistik, DBS) was provided by the Deutsches Bibliotheksinstitut up to 2001. In the future data will be collected by Hochschulbibliothekszentrum des Landes Nordrhein-Westfalen (HBZ). A brief introduction to German libraries is: Jürgen Seefeldt and Ludger Syré: *Portals to the Past and to the Future. Libraries in Germany*. Hildesheim : Olms, 2003.

12
CHAPTER

Managerial Convergence in Evolution Public Institutions in the European Union

Georgeta MODIGA[1]

Associate Professor. PhD, "Danubius" University of Galati, Faculty of Law, Romania

ABSTRACT

Although public management is a new management science, there are already convinced that the extension of the principles and specific approaches, not just a relative thing, it becomes imperative that determines the coordinates of major public sector reform. Otherwise there is the risk of slipping into a formalism stressed, changing some generally accepted fundamental and essential to the detriment of meeting the new management public. Consequently, it is absolutely necessary to give the perception and treatment of the old administrative system in general and public institutions in particular as bureaucratic administrative apparatus are developed rules, regulations, and laws by which they are applied and the transition to new general management principles and legitimate public that the administrative system as a whole and each public by public managers seeking a specified level of performance management, reflected in increasing public interest and satisfaction of specific social needs. Communication is one of effective management tools of change in the organization. It can help to adjust the attitude of the "way of looking at things" and to change behaviors. Mission and objectives managerial communication are closely related to organizational changes and environmental characteristics in which the organization functions. Public relations are essentially communication activities if relationships public management of communication between an organization and its public interest based on public, effective public relations manager must be in permanent contact with the public organization to be able to differentiate any time their communication needs, formulate and convey the message to the characteristics of each of them and track their response to receiving each of the messages.

Keywords: public institution; public management; public relations

Over the years, management in some organizations is becoming more distinct and sophisticated at the same time organizations become much stronger and longer lasting. It is enough to recall the Roman Empire that lasted hundreds of years, where legions of the peoples of Europe and the Middle East, much less organized and coordinated. New lands were known to the responsible government of Rome and roads were built to cross the entire empire, the information and follow them to be relieved and so center empire to intervene promptly and effectively in times that problem.

This communication, as we shall see, is an essential requirement for successful management. Roads so famous, some of which are used today, at the same time helped to quickly raise taxes, and the development and transmission of real-time decision for solving problems facing the empire. Of course the importance of these roads have been much higher in legions able to be very mobile, can move easily from one corner of the empire to another to intervene where needed in case of riot or other events.

History shows that, although the organization may be regarded as old as humanity, before the twentieth century can hardly find ways systematically designed and developed the model in which they are conducted. People were especially interested only in how organizations can be used to obtain money or political power

European Union has triggered the formation of a new body of officials in the European landscape, whose legal status has specific elements of both the civil servants at national and international officials to. Since the first time that the European Union was founded, it was necessary to regulate the legal status of officials of the institutions. Over time, the body of European civil servants underwent several changes both in terms of number and in terms of organization.

Management is a complex activity, with profound implications for the economic, social, political witnessed throughout its evolution and accept different approaches. Such management can be seen as a "process of coordination of human resources, information, - Physical and Financial achievements regarding organization goals." According to his opinion A. Mackensie, management is "a process in which the manager operates with three basic elements: ideas, things and people, making objective through others."

The new public management is met as "market-oriented government" (market based public administration), "post bureaucratic paradigm","entrepreneurial government" (Osborne and Gaebler, 1993). Each of these elements contains relatively similar names:

1) the major change from traditional public administration to focus on results and personal accountability of managers in public administration;

2) from the way traditional organizational - bureaucratic organization of public administration through flexible personnel policies, terms and conditions of employment;

3) performance measurement (performance measurement) in government activities through performance indicators (performance indicators) - systematic evaluation of programs based on the three E: saving, efficiency and effectiveness;

4) the top management of the organization is more committed political decision-makers are at a time in the management of public organization;

5) the government provided public services are exposed to market mechanisms through mechanisms external contracting (contracting out) - separating routing orientation (separation of steering of rowing) (see also Osborne & Gaebler, 1993);

6) reduced government provided public services through privatization and subcontracting (McKinney & Howard, 1998).

Measures to shift from traditional to new public management administration proposed an amendment to NPM are appropriate emphasis in public administration activities from process to results. These are the changes in personnel management, budgeting mode - program budgeting (budget based schedule), performance- oriented budgeting (performance based budget), accounting results (accrual accounting).

Outline of the new management guidelines focus largely on achieving results, objectives and empowering managers to achieve these goals, managers take political decisions, not just technical performance measurement through clear performance indicators, the introduction of competition in the public sector government and less implementation (steering Rather than rowing), privatization of public enterprises.

Traditional Administration owes the theoretical paradigm of Weberian bureaucracy and the notion that politics should be separated from the technical, civil service (civil service) should, according to this view, to be independent, neutral with political objectives, borrowing Weberian concept of neutrality axiological, social sciences over and positivism (sociologist is a social engineer, has a deal with the goals, but only at the means).

The theoretical basis of new public management guidelines are neo -classical economics and private management, rational choice theory, models public - choice follows as a consequence, the assumptions of neo -classical economics, namely the development of its models within the public sector. Social elections at all levels are considered normal, data, therefore public sector employees make choices, their responsibilities are not just technical agencies have established policies at other levels.

The new public management reform directions thus summarizes: disaggregating, competition, boosting performance. On the other hand, Lane (2000) identifies the characteristic of contraction as a basis for the new public management authority as opposed to primarily used for traditional administration. In addition the new public management tends to replace long-term contracts specific to management (for example, human resources, contracts are indefinitely officials) with short-term contracts (similar, though not so " short " contracts governing the exchange of goods and services - "on the spot" or "spot market contracting") with specific, measurable, specific contracts market and corporate governance. This has important implications in terms of competition, allowing the introduction of market-specific mechanisms in the public sector. Contracting refers to both the external contracting of public services (contracting -out), egg achieving a highway contracting with public money following a public auction, and the contraction in the public sector. If the latter alone contracting agency with higher authority setting targets and performance indicators, and the contracting staff usually at managerial levels.

Professional Management - Managers have responsibility for the results, but also the authority to make decisions, often political, not just technical assume decisions have views they support the public, not just technicians apply policies to others. They are sworn in under a management plan with objectives and performance indicators (level of government), public assumes that plan and figure out how to implement it.

They act like a CEO (chief executive officer) for private companies accountable to the Board of Directors who sworn in and, finally, the general meeting of shareholders. Professional managers who are accountable to the authority vested in office, but they are but agencies and citizens as such, although not directly elected by them, have the freedom to tell citizens through the media, of course policies they promote. By contrast, remember that officials are just technicians; they cannot sustain political viewpoints while on duty and cannot have position papers on public policies they implement technically.

Data refer to measuring progress, operationally implementation, while standards set by comparison desirable levels (minimal or excellence) to be achieved. Management reform emphasized performance levels of public officials, be they offices, government agencies or employees. In all areas, at all levels are required the development of performance indicators to assess fail informal and collegial organization or individual progress towards goals asumate.5

- Control outputs rather than procedures. The focus is on control results, agencies having greater freedom to plan the budget, human resources, strategy so effectively monitor the results. For example, if schools in Romania, the employment of human resources are done by the inspectorate, leaving very little freedom of schools to promote their policies and actively monitoring results through employment and wage policies. School education reform school management provides the ability to contract directly needed by teachers according to the objectives and needs of each school.

- Disintegration and fragmentation of the public service agency. Disaggregating means dividing broad ministerial department's smaller agencies that operate under managerial autonomy, each being responsible for the provision of well-defined on the underlying domain policies. The model was that of the Next Steps agencies introduced in 1988 in the UK Thatcher government. The basic idea is summarized in the government report recommending the establishment of such agencies "should be established in order to carry out the executive functions of government in a policy and a budget established by a ministerial department." It is assumed that such agencies will become fragmented into a contractual or quasi -contractual relationship (the setting of "target" levels of achievement) with hierarchical authority (department / ministry) for the provision of clear, with good quantity and quality standards specified introducing flexibility and competition in the public sector.

In addition, their employees can be largely contracted staff in turn attached with explicit goals and performance levels.

- The public sector works as an ethics that is different to serving the public interest, the citizen and not the provision of services to certain customers, clients and citizens' interests may not coincide at one point or the public sector has a duty to serve the interest citizens that is both more general and more diffuse cannot always be accurately defined by indicators and targets;

- Specify the public sector (the impossibility of establishing clear objectives and performance indicators - profitability cannot be a criterion citizen is not just a client, introducing market mechanisms do not work in all cases). "Public management cannot be just a simple economical way to maximize customer satisfaction "(McKinney & Howard, 1998). The public sector ought not only that: he must reconcile a multitude of values and interests, to redistribute resources according to equitable principles and enforce constitutional values.

- Public sector ethos (lifetime employment, the procedures, wage depending on the position, and loyalty to the system) has been altered by the introduction of performance-based monetary systems that often lead to perverse incentives - individuation, loss spirit of teamwork, collective responsibility

Accountability in the public sector is mediated by politicians, officials and politicians are accountable to, and they to the electorate. On the other hand, the higher authority (ministerial department) has the capacity (human resources, expertise) monitor work of agencies, making their political responsibility to be poor; rather they are evaluated by customers as service provider. However, the integration of services into a coherent program of government fades become quasi - autonomous agencies and accountability to politicians and citizens is diluted.

New Public Management is part of the first generation of public sector reforms, being guided by its efficiency, in particular by reducing costs, yet the public can not be guided only by increasing the quantity of products (outputs) services, but should be focused on quality of life than increasing the quantity and quality of service.

Improved quality of life involves two issues: first, it is an integrated approach to social problems that transgresses simple lines delineated the various services (excessive concern for outputs - products - at the expense of results - outcomes - changes in conditions, behaviors, attitudes, consequences of implementing programs) and secondly the question distribution, equitable allocation of services. Both issues have tended to be ignored by the new public management particularly concerned strictly economic efficiency. An example may be the Romanian higher education where funding is granted, for the most part, on the student equivalent staff.

However, this is only one output, not directly linked to improved public service for citizens. Such a result would be to increase the employability of graduates, their effectiveness (achieved wage) and even social benefits to society (inventions, scientific discoveries, increased productivity associated knowledge, etc.)

Restructuring the public service increased institutional complexity and decreased autonomous capabilities (skills) citizens in solving public problems (multiplication agencies confuses citizens do not know where to go), which led to the failure of the new public management in solving social problems.

Disintegration was followed by reinsertion of departmental control services for government policy to be pursued. The competition was held, but indirectly by measuring performance and setting tables, while stimulation is maintained, but in turn creates problems related to increasing inequalities in the public and uneven performance.

Some specific trends NPM reform stalled, even some processes become reversible after it was found that long-awaited efficiency led to weakening public position and the impossibility of implementing coherent policies to citizens. Reverse trends are manifested by:

- Reinsertion of public sector activities in the private sector had contracted (outsourced) - re- nationalization of the company in charge of rail infrastructure in the UK and airport security in the United States after the attacks of September 11.
- Another trend was that, contrary to the fragmentation (or disaggregating), the reunification of the various agencies to promote coherent policies, and this time the example of the Department of Homeland Security in the United States emerged as a result of the September 11 attacks by unifying approximately 22 federal agencies that aimed fragmented different aspects of internal security. Fragmentation proposed new public management has led to duplication of many hierarchs, to inefficiency and inability coherent policy; example rail regulatory agencies in the UK, the number 3, covering areas fragmented (balkanization, bureaucrats boutique) Railway Safety, licensing railway companies and the investment in rail infrastructure.

All these changes mean moving the focus from business -type efficiency of the administrative unit (which led to the disintegration and fragmentation) on providing quality services citizens, clients of public programs. Particularly effective results in economy and competitiveness achieved through lower costs, however, fragmentation has led to the creation of artificial public service areas, the "balkanization" of bureaucrat's services and the emergence of "boutique" that institutionalize their fields, and their functions without their necessarily a real necessity came from the company in respect of those matters. Unification agencies and holistic approach to many public services becomes possible by using the new media in the context of government information. Modern communication media influence on public management is achieved through a wide range of changes in cognitive, behavioral, organizational, political and cultural issues related to information systems - the digital era governance.

Digital era governance proposed, instead, a focus on customer needs and eliminating red tape, not so much by introducing competition as using information resources in the provision of public services. Thus, there are services that save time and facilitate customers like "one stop". The provision of such "one-stop" takes various forms, including "one-stop shops" ("one -stop agencies"- multiple administrative services are provided by staff co- located), "unique web windows "(a client-centered integrated interface where multiple services are integrated electronically). Also, such services require greater transparency of public management processes.

New Public Management could not determine governance agility to adapt to new global challenges, to refresh their structures to respond to changing needs and volatile external environment. He remained concerned pre Efficiency and cost reduction, losing sight of the very specific public sector and governance mechanisms for efficient techniques that cannot be substituted but complex processes responsible for implementing collective decisions.

An important aspect of public administration reform is the reform management in the application. A major risk in public administration reform is clutter everyday tasks, leaving little time to look for new solutions to malfunctioning occur.

Consequently, for public administration reform to be successful it is necessary that a large number of target groups (in particular key people at management levels and decision) to express their support and commitment in recognizing the need for change and the implementation of these changes.

Based on the model of integrated development organizations, we can identify four factors blocking the implementation of the updated strategy: strategic (decisional complexity), structural (specify any system bureaucracy, limited human and financial resources, size and complexity), cultural (fear of risk tradition of "continuity" change of mentality) and behavior (lack of individual incentives, misunderstanding ultimate goals, motivation and anxiety behaviors tunes).

Public administration cannot be reformed in a short time. It is a long term process that must be implemented by several consecutive governments in a difficult environment, competitive and constantly changing. Therefore, it is needed to obtain a political consensus regarding the requirements necessary to achieve the whole process of reform by:

- Large scale dissemination of information on the reform and encouraging citizen participation, civil society, the business environment, the public servants to reform through a democratic exchange of views about the development process;
- Funding reform based on a professional analysis of the current state of the activity and performance of the public administration and based on periodic assessment of the consequences of activities completed;
- Using public administration reform experiences of other countries, especially EU Member States and candidate

countries, while maintaining historical and cultural elements of the Romanian public administration;

- A comprehensive reform approach: no change should not be isolated or partially realized, if it was not conceived as an integral and organic part of the whole reform strategies;
- An open approach to reform: reform strategy components will be updated and adapted to changes in the external environment, but other components of reform and transformation, to use in the best way during implementation experience;
- Establish strategic priorities: for each step is set a limited number of priority targets on which will point toward and be allocated the necessary funds. They must be those changes that advance the reform as a whole and shall determine or influence all other transformations;
- Ensure continuity of public administration officials, so that their function during reorganization, decentralization, and so on;
- Coordination and management of the entire reform from a single point at the highest level of government, to ensure coherent implementation strategy. In the reform process, the main factors involved in the coordination, implementation and monitoring measures envisaged are:
    - Prime Minister will have the role of coordinating the overall government reform and delegate specific tasks to all Ministers responsible;
    - The whole process of public administration reform will be coordinated by the Minister of Administration and Interior;

- Each minister must submit a statement of intent on strategic guidelines for ministry he leads, which may be updated periodically;
- Superior Council for Public Administration Reform, Public Policy Coordination and structural adjustment will analyze sectorial strategies for each ministry or in whole, and make suggestions and recommendations to ministers by the Prime Minister;
- Three ministers have a special role in the reform process: Minister of Interior, Minister of Public Finance and coordinating minister for the General Secretariat of the Government. The difference between the ministries headed by three ministers in relation to other ministries is that they have a general responsibility for key aspects of the whole public administration, and a specific responsibility for the sector they represent;
- The Ministry of Interior actors involved in the reform process under the authority of the Minister are:

- Central Unit for Public Administration Reform, Director of the Ministry of Interior, provides technical support and monitors the entire process of reform. It also provides structure with the Public Policy Unit of the Government Council Secretariat, implementing its decisions;
- National Institute of Directors together with the regional centers is responsible for the training of civil servants in central and local public administration; National Agency of Civil Servants is responsible for the management of the public.¬

- Monitoring Groups PAR public counties and ministries coordinated by the Central Unit for Public Administration

Reform. These groups are a national network of modernizers, representing an interface between the central and local public administration, between the political and the technical apparatus of public administration.

- Ministry of Finance should develop a new budget system oriented programs and performance to support effective management of performance across the public administration, budget execution processes more flexible and transparent accounting techniques allowing both accounting for assets, as and budgetary flows.

- SGG must prepare systematic reform, long-term management policymaking and policy analysis and evaluation of the central decision-making process;

- In the General Secretariat of the Government, was recently established Public Policy Unit whose mission is to strengthen the capacity of government to coordinate the formulation, implementation and monitoring of public policies, ensuring the role of interface between different institutions with a decisive role in the formulation of public policies.

The concept of performance monitoring plan is one of the most important aspects of public policy reform, here was also included on decentralization policies and the management of the public

Anticorruption policies will be developed taking into account the recommendations of GRECO and Anti-Corruption Manual issued by the United Nations in 2003. They will be applied based on the following conditions and principles:

- institutional assessment, to identify, define and correctly apply measures to fight corruption and not just chaotic to introduce new regulations, external pressure;

- ensuring the political neutrality of applying anticorruption measures, and the establishment of independent monitoring and evaluation mechanisms;
- transparency of public authorities in developing policies and their application process, including unfettered access to public information;
- achieving the partnership with civil society, the development of institutional mechanisms that ensure the widest possible participation of the citizens in assessing the extent of corruption that affect them directly, and through participation in civic organizations;
- integrated approach by defining the exact sources that generate corruption and coordinating policies and developing common mechanisms to ensure cooperation of public authorities and competent clear demarcation between them, instead of confusion as there is currently competent and that hinder effective implementation applicable laws;
- results orientation by introducing mechanisms for monitoring and evaluation.

REFERENCES

1. Andronicescu, Armenia (2005). New Public Management. Bucharest: University Publishing House.
2. Andronicescu, Armenia (2005). Public Management. Case Studies of Institutions and Government Authorities. Bucharest: University Publishing House.

3. Matthew, L. (2002). Management of Local Development - Decentralization. Innovation. Risk. Bucharest: Economic Publishing House.

4. Olaru, S. D. & Sun, C. R. (2001). Management Public Relations and Anger Management. Bucharest: Lumina Lex Publishing House.

5. Paunescu, Michael (2008). Public Management in Romania. Iasi: Polirom.

6. Profinoiu, M. (2009). Strategic Management of Local Communities. Bucharest: Publishing House.

7. Smith, I. D. (2003). Management. Galati: Academic Publishing House "Danubius".

Index

S

T